Dynamic Programming and
Partial Differential Equations

This is Volume 88 in
MATHEMATICS IN SCIENCE AND ENGINEERING
A series of monographs and textbooks
Edited by RICHARD BELLMAN, *University of Southern California*

A partial list of the books in this series appears at the end of this volume. A complete listing is available from the Publisher upon request.

Dynamic Programming and Partial Differential Equations

EDWARD ANGEL

Department of Electrical Engineering
University of Southern California
Los Angeles, California

RICHARD BELLMAN

Departments of Mathematics and Engineering
University of Southern California
Los Angeles, California

ACADEMIC PRESS New York and London 1972

ACADEMIC PRESS, INC.
111 Fifth Avenue, New York, New York 10003

United Kingdom Edition published by
ACADEMIC PRESS, INC. (LONDON) LTD.
24/28 Oval Road, London NW1 7DD

LIBRARY OF CONGRESS CATALOG CARD NUMBER: 70-189674

AMS(MOS) 1970 Subject Classifications: 35J60, 35K55

PRINTED IN THE UNITED STATES OF AMERICA

Contents

CHAPTER 3

DYNAMIC PROGRAMMING

CHAPTER 4

THE POTENTIAL EQUATION

CHAPTER 5

DYNAMIC PROGRAMMING AND ELLIPTIC EQUATIONS

CHAPTER 6

INVARIANT IMBEDDING

CHAPTER 7

IRREGULAR REGIONS

CHAPTER 8

SPECIAL COMPUTATIONAL METHODS

CHAPTER 9

UNCONVENTIONAL DIFFERENCE METHODS

CHAPTER 10
PARABOLIC EQUATIONS

CHAPTER 11
NONLINEAR EQUATIONS AND QUASILINEARIZATION

APPENDIX
COMPUTER PROGRAMS

Preface

A major task of contemporary mathematical analysis is that of providing computer algorithms for the numerical solution of partial differential equations of all types. This task requires a motley mixture of methods. In this book we wish to show first that dynamic programming and invariant imbedding furnish some powerful techniques for the solution of linear elliptic and parabolic partial differential equations over regular and irregular regions. Nonlinear equations are made accessible by the use of quasilinearization which can be combined with the earlier procedures to study identification and inverse problems. What is quite important is that these algorithms are straightforward, easily learned, readily programmed, and easily used.

We wish to thank John Casti, Dave Collins, Nestor Distefano, and Art Lew for much helpful advice and comments in the preparation of the manuscript. We are particularly grateful to John Todd of the California Institute of Technology who read the manuscript through in his usual painstaking fashion and made a large number of suggestions based on his vast erudition and considerable experience in this field which materially improved both content and presentation. Edward Angel was with the Department of Electrical Engineering and Computer Sciences, University of California at Berkeley during the preparation of the manuscript.

Chapter I

Introduction

The principal aim of this volume is to apply some of the ideas and methods of the theory of dynamic programming to the analysis and computational solution of the potential equation

$$u_{xx} + u_{yy} = 0\,, \tag{1}$$

with the boundary condition

$$u(x, y) = g(x, y), \qquad (x, y) \in \Gamma\,, \tag{2}$$

where Γ is the boundary of a region R which may be of quite irregular shape as in Fig. 1.

FIGURE 1

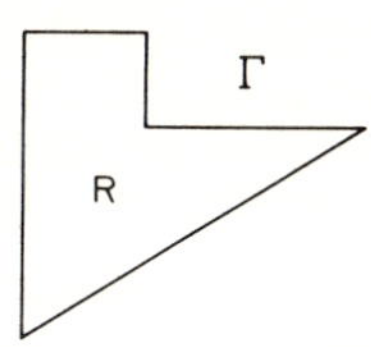

There are currently a large number of methods available for handling questions of this nature, all of which work well in carefully chosen circumstances. All, however, possess various drawbacks, including, of course, the methods presented below. It is this fact which motivates a constant search for new methods. The pros and cons of these techniques are discussed in appropriate places.

Variations on the principal method and the principal equation lead to other methods and equations of interest in their own right. We shall discuss these below. Let us begin with the basic idea.

In variational problems involving functions of one space variable, say the minimization of

$$J(u) = \int_0^T g(u, u')\, dt\,, \tag{3}$$

subject to $u(0) = c$, a fundamental observation is that the minimum value of $J(u)$ depends upon the quantities c and T. We then regard c and T as variables, with $T \geqslant 0$ and $-\infty < c < \infty$, and write

$$f(c, T) = \min_u J(u)\,. \tag{4}$$

The original variational problem is thus transformed into that of obtaining an equation for $f(c, T)$. This is the approach of the theory of dynamic programming: we imbed a particular problem, with fixed values of c and T, in a family of problems in which c and T assume ranges of values. Then we obtain relations connecting various members of the family.

An equation for $f(c, T)$ is obtained by viewing the minimization process as a multistage decision process in which a choice of a function over $[0, T]$ is decomposed into a choice over the subinterval $[0, s]$ and then a choice over the subinterval $[s, T]$, $0 < s < T$. Thus

$$\min_{u[0,T]} = \min_{u[0,s]} \min_{u[s,T]}\,. \tag{5}$$

This is readily translated into a functional equation for $f(c, T)$ using the principle of optimality. If we let $s \to 0$, this equation becomes a partial differential equation. This is discussed in Chapter 3.

A similar approach can be pursued in the multidimensional case. We begin with the fact that the partial differential equation in (1) is the Euler–Lagrange equation associated with the quadratic functional,

$$J(u) = \int_R (u_x^2 + u_y^2)\, dR\,, \tag{6}$$

the Dirichlet functional. As above, we can regard the minimization of $J(u)$ as a multistage decision process in which the choice of a function over R is decomposed into a choice over a subregion R_1 and then over the region $R - R_1$. This yields

$$\min_{u \in R} = \min_{u \in R_1} \min_{u \in (R - R_1)} . \tag{7}$$

Different choices of R_1 yield a variety of interesting and useful results. Thus, for example, if the region R is rectangular, we can choose R_1 as a rectangular strip (Fig. 2). This leads to either a recurrence relation or a

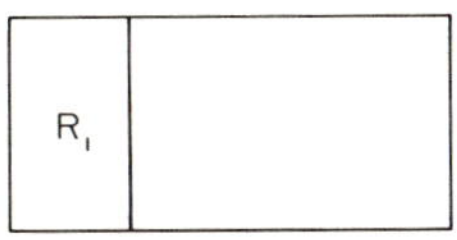

FIGURE 2

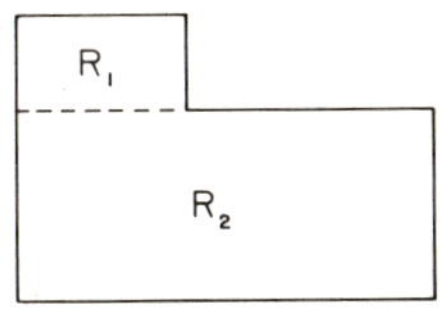

FIGURE 3

differential equation depending upon the type of discretization we employ and whether R_1 is taken to be infinitesimal or not. If the region has the form shown in Fig. 3, we choose R_1 as indicated in the figure and the equation over the region $R = R_1 + R_2$ can be solved in terms of its solutions over the simpler regions R_1 and R_2.

All of this will be discussed in some detail in subsequent chapters, and numerical applications will be given. We shall not, however, discuss the corresponding problems and procedures for circular and spherical regions.

Let us now examine the contents of individual chapters. In Chapter 2 we review the minimization of the quadratic functional

$$J(x) = \int_0^T [(x', x') + (x, A(t)x)]\, dt, \tag{8}$$

using the classical approach of the calculus of variations. This leads to the Euler–Lagrange equation

$$x'' - A(t)x = 0, \tag{9}$$

subject to various boundary conditions, depending on the conditions originally imposed on x. Some of the problems encountered in a numerical solution are discussed there. At the close of the chapter there is a brief sketch of the Rayleigh–Ritz and Bubnov–Galerkin methods. These are powerful techniques for treating the linear partial differential equations of mathematical physics and engineering. The drawback is that they lead to the task of solving large systems of linear algebraic equations.

Our concern with questions of this nature is due to the fact that discretizations of various types transform the Dirichlet functional into a functional of the foregoing type and thus to equations of the type appearing in (9).

In Chapter 3 we present the dynamic programming approach to the minimization of $J(x)$ as well as to the minimization of its discrete version

$$\sum_{n=0}^{N} [(x_{n+1}-x_n, x_{n+1}-x_n) + (x_n, Ax_n)] . \tag{10}$$

In this way we encounter the Riccati equation and its discrete analogue.

Chapter 4 is devoted to the potential equation, Green's functions, and various discretization procedures. Chapter 5 contains a more detailed discussion of the application of dynamic programming to the discrete case.

Chapter 6 describes some applications of the theory of invariant imbedding to the solution of linear partial differential equations. An advantage of this technique is that we can apply it to general linear equations, not necessarily derived from a variational principle.

In Chapter 7 we discuss the details of the application of the foregoing methods to irregular regions, while Chapter 8 is devoted to some special methods which can be used for regular regions.

The theory of dynamic programming yields some nonlinear partial differential equations describing continuous decision processes which are conveniently resolved computationally using the analogous equations for the associated discrete decision processes. This suggests a corresponding procedure for other types of partial differential equations. To illustrate this idea, in Chapter 9 we examine the use of the recurrence relation

$$v(x,t+\Delta) = v(x+v(x,t)\Delta,t), \qquad t = 0, \Delta, 2\Delta, \ldots, \tag{11}$$

to study the equation

$$u_t = uu_x, \tag{12}$$

and some associated questions.

The parabolic equation

$$u_t = u_{xx} + u_{yy}, \tag{13}$$

the heat equation, is studied in Chapter 10 using the Laplace transform to reduce it to an equation of elliptic type which is then studied by the foregoing methods. Numerical inversion of the Laplace transform then yields values for u.

Finally in Chapter 11 we examine the use of the methods of successive approximations, using quasilinearization, to resolve nonlinear equations.

Throughout we are concerned both with the analysis itself which is extremely interesting and with the equally interesting question of providing numerical answers to numerical questions using reasonably simple straightforward methods and easily written computer programs. Our constant theme is that these two types of problems are closely interrelated. The analysis must be judged on whether or not it is computationally operational, while the derivation of operational methods often require new analytic procedures.

Chapter 2

Quadratic Variational Problems

1. Introduction

In this chapter we wish to review some of the more important classical methods for treating linear differential equations and quadratic variational problems. An understanding of what is required will motivate our subsequent application of a different approach, dynamic programming to the same problems. At the outset however, it should be explicitly pointed out that all methods face difficulties in treating problems of high dimension. These difficulties are different for different methods.

We shall first discuss the minimization of the scalar quadratic functional

$$J(u) = \int_0^T (u'^2 + q(t)u^2)\,dt\,, \tag{1}$$

subject to $u(0) = c_1$, $u(T) = c_2$ and then the N-dimensional counterpart, the problem of minimizing

$$J(x) = \int_0^T [(x', x') + (x, A(t)x)]\,dt\,, \tag{2}$$

subject to $x(0)=c$, $x(T)=d$. Our tool will be the associated Euler–Lagrange equation.

Following this, we shall briefly sketch the Rayleigh–Ritz and Bubnov–Galerkin methods. All three procedures ultimately require the solution of linear algebraic equations, and the obstacles stem from this.

2. Variational Approach

Let us suppose that there is a function u such that u and u' belong to $L^2(0,T)$, u satisfies the boundary conditions, and u furnishes the absolute minimum of $J(u)$. Our aim is to obtain a necessary condition that u must satisfy. This is the Euler–Lagrange equation which in this case turns out to be sufficient.

We proceed as follows. Let v be another function such that v and v' belong to $L^2(0,T)$ with $v(0)=v(T)=0$. Then for any real ϵ, the function $u+\epsilon v$ satisfies the original boundary conditions and is such that it and its derivative are in $L^2(0,T)$. Consider next the expression

$$J(u+\epsilon v) = J(u) + \epsilon^2 J(v) + 2\epsilon \int_0^T [u'v' + q(t)\,uv]\,dt, \tag{1}$$

which by assumption possesses an absolute minimum at $\epsilon = 0$. This fact yields the variational condition

$$\int_0^T [u'v' + q(t)\,uv]\,dt = 0 \tag{2}$$

for all v of the foregoing nature.

Integrating by parts, this becomes

$$v\int_0^t q(t_1)\,u(t_1)\,dt_1\Big]_0^T + \int_0^T \left[u'v' - v'\int_0^t q(t_1)\,u(t_1)\,dt_1\right] dt = 0\,. \tag{3}$$

The integrated term drops out since $v(0)=v(T)=0$. Hence, for any constant c_3,

$$\int_0^T v'\left[c_3 + u' - \int_0^t q(t_1)\,u(t_1)\,dt_1\right] dt_1 = 0\,. \tag{4}$$

Choose

$$v' = c_3 + u' - \int_0^t q(t_1)\,u(t_1)\,dt, \qquad v(0) = 0, \tag{5}$$

where c_3 is determined by the condition $v(T)=0$. Then (4) yields the relation

$$c_3 + u' - \int_0^t q(t)_1 u(t_1)\,dt_1 = 0 \tag{6}$$

almost everywhere. The almost everywhere may be replaced by everywhere. Thus (6) yields the Euler–Lagrange equation

$$u'' - q(t)u = 0, \qquad u(0) = c_1, \qquad u(T) = c_2. \tag{7}$$

3. Positive Definiteness, Existence and Uniqueness of Solution

It is not difficult to show that if $J(u)$ is positive definite, then (2.7) possesses a unique solution. We proceed as follows. If there are two solutions, there is a solution v such that $v(0)=v(T)=0$. Consider then the expression

$$\begin{aligned} 0 = \int_0^T v(v'' - q(t)v)\,dt &= vv'\Big]_0^T - \int_0^T [v'^2 + q(t)v^2]\,dt \\ &= 0 - \int_0^T [v'^2 + q(t)v^2]\,dt, \end{aligned} \tag{1}$$

a contradiction to the assumed positive-definite character of $J(v)$ if v is not identically zero.

That (2.7) possesses a solution follows from the discussion below.

4. Computational Aspects

Let u_1 and u_2 be the principal solutions of (2.7), i.e.,

$$\begin{aligned} u_1(0) &= 1, & u_2(0) &= 0, \\ u_1'(0) &= 0, & u_2'(0) &= 1. \end{aligned} \tag{1}$$

Write

$$u = a_1 u_1 + a_2 u_2, \tag{2}$$

where a_1 and a_2 are to be determined by the conditions

$$c_1 = a_1, \qquad c_2 = a_1 u_1(T) + a_2 u_2(T). \tag{3}$$

The positive-definite nature of $J(u)$ guarantees that $u_2(T) \neq 0$, as indicated in Section 3, whence a_1 and a_2 are uniquely determined. The values of the principal solutions are determined by numerical integration of the differential equation (2.7) with the appropriate initial conditions.

5. Vector–Matrix Case

Let us consider in similar fashion the task of minimizing

$$J(x) = \int_0^T [(x', x') + (x, A(t)x)]\, dt, \tag{1}$$

where $A(t)$ is positive definite. We suppose that $x(0) = c$, $x(T) = d$ and $x' \in L^2(0, T)$. Then proceeding as before the variational equation is

$$x'' - A(t)x = 0, \qquad x(0) = c, \quad x(T) = d. \tag{2}$$

Analogous arguments to those given in Section 3 show that this equation possesses a unique solution.

The computational aspects, however, require a careful examination. As a straightforward extension of the method used in the scalar case, write

$$x = X_1(t)a + X_2(t)b, \tag{3}$$

where X_1 and X_2 are the principal matrix solutions of

$$X'' - A(t)X = 0, \tag{4}$$

that is,

$$\begin{aligned} X_1(0) &= I, \qquad X_1'(0) = 0, \\ X_2(0) &= 0, \qquad X_2'(0) = I, \end{aligned} \tag{5}$$

and a and b are constant vectors to be determined by the boundary conditions given in (2). These lead to the equations

$$c = a, \qquad d = X_1(T)a + X_2(T)b. \tag{6}$$

It is not difficult to extend the method sketched in Section 2 to the present case to show that $X_2(T)$ is not singular. Hence a and b, and thus $x(t)$, are uniquely determined by (2). This determination of b, however, requires a solution of a system of linear algebraic equations, always a ticklish matter. It is particularly so when the dimension of a is large and T is large.

The fact that T is large means that $X_2(T)$ is close to a singular matrix. This combined with the fact that $\dim(X_2)$ is large means that numerical accuracy is not readily ensured. A detailed discussion of these matters will be found in the reference cited in the bibliography at the end of the chapter.

It is the block to a straightforward solution of linear differential equations of high dimension that motivates the constant search for new approaches to linear partial differential equations and quadratic variational problems.

6. Rayleigh–Ritz Method

Let us briefly sketch two of the most powerful approaches which avoid the route indicated in Section 5. The first is the Rayleigh–Ritz method. Consider the functional

$$J(x) = \int_0^T [(x', x') + (x, A(t)x)]\, dt, \tag{1}$$

with $x(0) = c$, $x(T) = d$, and use the trial function

$$x = \sum_{k=1}^{M} b_k \phi_k(t). \tag{2}$$

We have a choice of letting the ϕ_k be scalars and the b_k vectors or conversely. In any case, let us choose the ϕ_k as known functions and the b_k as unknowns, subject only to the boundary conditions.

Then

$$J(x) = J(b_1, b_2, \ldots, b_M) \tag{3}$$

and the minimization is now over the b_k. This leads to a set of linear algebraic equations, of degree M if the b_i are scalars. If $M \ll N$, this is a much more tractable problem than the original.

If we do not wish to use the boundary conditions, we can form the new functional

$$J(x, \lambda_1, \lambda_2) = J(x) + \lambda_1(x(0) - c, x(0) - c) + \lambda_2(x(T) - d, x(T) - d), \tag{4}$$

where λ_1 and λ_2 are Courant parameters and λ_1 and $\lambda_2 \gg 1$.

The choice of the sequence $\{\phi_k(t)\}$ is critical and is determined in any particular case by some combination of mathematical reasoning, physical intuition, and experience.

7. Bubnov–Galerkin Method

In place of solving the Euler–Lagrange equation

$$x'' - A(t)x = 0, \qquad x(0) = c, \quad x(T) = d, \tag{1}$$

consider the problem of minimizing

$$J_1(x) = \int_0^T (x'' - A(t)x, x'' - A(t)x)\, dt \tag{2}$$

over $x(t)$ such that the boundary conditions are fulfilled and the integral exists. Now use the Rayleigh–Ritz method to carry out this minimization.

It is sometimes convenient to introduce the mixed expression

$$J_2(x) = J(x) + \lambda J_1(x) \tag{3}$$

in place of $J(x)$ alone, where λ again is a Courant parameter.

BIBLIOGRAPHY AND COMMENT

Sections 1–5. A detailed discussion of these problems plus a number of additional references may be found in

R. Bellman, *Introduction to the Mathematical Theory of Control Processes, I: Linear Equations and Quadratic Criteria*, Academic Press, New York, 1967.

For an introduction to the regularization techniques of Tychonov, see

R. Bellman, R. Kalaba, and J. Lockett, *Numerical Inversion of the Laplace Transform*, Amer. Elsevier, New York, 1966.

Sections 6–7. See

R. Bellman, *Methods of Nonlinear Analysis*, Vol. 1, Academic Press, New York, 1970.

Chapter 3

Dynamic Programming

1. Introduction

In this chapter we wish to describe some applications of the theory of dynamic programming to variational processes governed by either differential or difference equations. The techniques yield particularly elegant results when applied to quadratic functionals associated with linear equations of state. Furthermore, and fortunately, these are the functionals and equations met in a variety of situations. Utilization of these results for numerical purposes will be discussed in the following chapters.

2. Difference Equations

Let x_n, $n = 0, 1, 2, \ldots$ be a K-dimensional vector, the *state vector*, determined recurrently by the difference equation

$$x_{n+1} = g(x_n, y_n), \qquad x_0 = c, \tag{1}$$

where y_n is an L-dimensional vector, the *control vector*. In many situations $L = K$. We suppose the set of control vectors, $\{y_n\}$, is to be chosen to minimize the *criterion function* (or "return" function),

$$J(\{x_n\}, \{y_n\}) = \sum_{n=0}^{N} h(x_n, y_n). \tag{2}$$

This defines a deterministic control process of discrete type. Assuming that the problem is well-posed in the sense that the minimum is achieved, as it is easy to show in the cases of interest to us in what follows, we set

$$\min_{\{y_n\}} J(\{x_n\}, \{y_n\}) = f_N(c), \tag{3}$$

a function defined for all c, and $N = 0, 1, 2, \ldots$. We see that

$$f_0(c) = \min_y h(c, y), \tag{4}$$

a function which in many cases is readily determined.

This is a fundamental *imbedding*, an idea we shall discuss in more detail in the chapter on invariant imbedding. In order to resolve a particular problem where c and N are specified, we proceed to obtain a functional equation for $f_N(c)$ considered as a function of c and N.

3. Functional Equation

As noted above, the function $f_0(c)$ is given by

$$f_0(c) = \min_y h(c, y). \tag{1}$$

In view of this determination, we would like to obtain a recurrence relation, another difference equation, which connects $f_{N+1}(c)$ with $f_N(c)$. Theoretically this would lead to a constructive way of obtaining the sequence $f_N(c)$. Operationally, as we shall see, there are some additional points that remain. The form of the criterion function in (2.2), its additivity, allows this relation to be readily obtained.

We have

$$\begin{aligned} f_{N+1}(c) &= \min_{\substack{\{y_n\} \\ n=0,1,2,\ldots,N+1}} J(\{x_n\}, \{y_n\}) \\ &= \min_{y_0} \min_{\substack{\{y_n\} \\ n=1,2,\ldots,N+1}} J(\{x_n\}, \{y_n\}) \\ &= \min_{y_0} \min_{\{y_n\}} \left[h(x_0, y_0) + \sum_{n=1}^{N+1} h(x_n, y_n) \right] \\ &= \min_{y_0} \left[h(c, y_0) + \min_{\{y_n\}} \sum_{n=1}^{N+1} h(x_n, y_n) \right]. \end{aligned} \tag{2}$$

Since $x_0 = c$, and $x_1 = g(x_0, y_0) = g(c, y_0)$, this yields the quantity

$$\min_{y_0} [h(c, y_0) + f_N(g(c, y_0))],$$

upon recalling the definition of $\{f_N(c)\}$. Since y_0 is a dummy variable here, we can eliminate the subscript and use the simpler form

$$f_{N+1}(c) = \min_{y} [h(c, y) + f_N(g(c, y))], \qquad N = 0, 1, \dots. \tag{3}$$

This is the fundamental recurrence relation which determines both the sequence $\{f_N(c)\}$ and the control variables which minimize.

4. Principle of Optimality

The functional equation in (3.3) becomes immediate when we conceive of the process as a multistage decision process and apply the principle of optimality of the theory of dynamic programming.

Principle of Optimality. *An optimal policy has the property that whatever the initial state and initial decision are, the remaining decisions must constitute an optimal policy with regard to the state resulting from the first decision.*

Frequently the applicability of the principle is immediate, as is a proof by contradiction. This is, for example, the situation in the foregoing problem. In all cases, however, it is necessary to validate its use.

5. Nonstationary Case

In many cases of importance the functions appearing in the description of the criterion function are time-dependent which means that J has the form

$$J(\{x_n\}, \{y_n\}) = \sum_{n=0}^{N} h_n(x_n, y_n), \tag{1}$$

where the single-stage return depends upon the stage.

To handle problems of this nature using the same concepts, we reverse the direction of time and introduce the function

$$\min_{\{y_n\}} \sum_{n=k}^{N} h_n(x_n, y_n) = \phi_k(c), \tag{2}$$

defined for $k = 0, 1, 2, \ldots, N$, $-\infty < c < \infty$. Here N is fixed and k, the starting time, is variable. The function $\phi_N(c)$ is now readily obtained, namely

$$\phi_N(c) = \min_y h_N(c, y), \tag{3}$$

and the desired recurrence relation is

$$\phi_k(c) = \min_y [h_k(c, y) + \phi_{k+1}(g(c, y))], \qquad k = 0, 1, \ldots, N-1. \tag{4}$$

6. Quadratic Functions

The functional equations can be considerably simplified in the case where the criterion function is quadratic and the state equation is linear. Consider, for example, the problem of minimizing the quadratic expression

$$J(\{x_n\}, \{y_n\}) = \sum_{n=0}^{N} [(x_n, Ax_n) + (y_n, By_n)], \tag{1}$$

where x and y are related by the equation

$$x_{n+1} = x_n + Cy_n, \qquad x_0 = c.* \tag{2}$$

Let us suppose that A and B are positive definite so that the minimum exists. Writing as before

$$f_N(c) = \min_{\{y_n\}} J(\{x_n\}, \{y_n\}), \tag{3}$$

a function defined for $-\infty < c < \infty$, $N = 0, 1, 2, \ldots$. We have, following (3.3),

$$f_{N+1}(c) = \min_y [(c, Ac) + (y, By)] + f_N(c + Cy)], \qquad N = 0, 1, \ldots, \tag{4}$$

with

$$f_0(c) = \min_y [(c, Ac) + (y, By)] = (c, Ac). \tag{5}$$

It remains to simplify (4) and thus obtain some relations of more tractable analytic and computational nature. To this end we make strong use

---

* We are employing the usual inner product representation of quadratic forms. Thus $(a, b) = \sum_{i=1}^{N} a_i b_i$, where a_i and b_i are the ith components of a and b respectively, whence $(x, Ax) = \sum_{i,\ j=1}^{N} a_{ij} x_i x_j$. We use $x_1, x_2, \cdots, y_1, y_2, \cdots$ to denote a sequence of vectors above to avoid the necessity of a superscript notation.

of the fact that $f_N(c)$ is quadratic function of c,

$$f_N(c) = (c, Q_N c), \tag{6}$$

where the matrix Q_N is independent of c.

The quadratic nature of $f_N(c)$ is apparent from the linear structure of the usual variational equations. In any case it may readily be established inductively using (4) as the following argument indicates. Using (4) and (6), we have

$$(c, Q_{N+1} c) = \min_y [(c, Ac) + (y, By) + (c + Cy, Q_N(c + Cy))]. \tag{7}$$

The point is that the minimum on the right hand side, as well as the minimizing value of y, can readily be determined. The minimizing y is a linear function of c. Thus, the evaluation of the expression will yield another quadratic form in c. Equating coefficients yields a recurrence relation connecting Q_{N+1} with Q_N. We shall carry through the details in the following chapter in connection with the variational problem of particular interest to us.

EXERCISES

1. Consider the problem of minimizing $J(\{u_n\}) = \sum_{n=0}^{N} [(u_{n+1} - u_n)^2 + u_n^2]$, where $u_0 = c$. Show by a change of variable that $f_N(c) = \min_u J = r_N c^2$.

2. Using the procedure indicated above, obtain a recurrence relation connecting r_N with r_{N+1}.

3. Show in this way that $\lim_{N\to\infty} r_N$ exists and determine its value.

4. Show directly that $r_N < r_{N+1}$, that r_N is uniformly bounded, and hence that r_N converges as $N \to \infty$.

5. Consider the function $f(c) = \min_u \sum_{n=1}^{\infty} [u_n^2 + (u_n - u_{n-1})^2]$, $u_0 = c$. Show that $f(c) = \min_v [c^2 + (v - c)^2 + f(v)]$ and hence determine r_∞ directly.

6. If $f(0) = 0$ and $f(c)$ is analytic in c, does the functional equation determine $f(c)$ uniquely?

7. Show that $\min_x [(x, Ax) - 2(x, y)] = -(y, A^{-1} y)$.

7. Minimum Convolution

The same ideas may be employed to obtain some more general results. Introduce the function of two variables

$$\phi_N(c, d) = \min_y J(\{x_n\}, \{y_n\}), \tag{1}$$

where x_n is subject to two constraints

$$x_0 = c, \qquad x_N = d, \tag{2}$$

and x and y are related by the relation

$$x_{n+1} = h(x_n, y_n). \tag{3}$$

Write

$$\begin{aligned} \phi_{M+N}(c, d) &= \min_y \Big[\sum_{n=0}^{M+N} \Big] \\ &= \min \Big[\sum_{n=0}^{M} + \sum_{n=M+1}^{N} \Big]. \end{aligned} \tag{4}$$

Let us choose a value for x_M, say $x_M = z$. Then (2) yields

$$\phi_{M+N}(c, d) = \min_z [\phi_M(c, z) + \phi_N(z, d)]. \tag{5}$$

EXERCISES

1. What modifications are necessary if the single-stage return $g(x, y)$, depends on n?

2. Consider the quadratic case and use (5) to derive a recurrence relation for r_{M+N} in terms of r_M and r_N, where r_N is defined as in the exercises at the end of Section 6.

8. Acceleration of Calculation

In many cases we wish to determine $f_N(c)$ or $\phi_N(c)$ for a particular value of N only. If N is large, it may be much more efficient to use the relations in Section 7 rather than relation (3.3). Thus, for example, (7.5) yields

$$\phi_{2^{N+1}}(c, d) = \min_z [\phi_{2^N}(c, z) + \phi_{2^N}(z, d)], \tag{1}$$

providing a relatively rapid determination of $\phi_1, \phi_2, \phi_4, \phi_8, \ldots$.

To calculate $f_N(c)$, we can use the relation

$$f_{N+M}(c) = \min_z [\phi_M(c,z) + f_N(z)], \tag{2}$$

and thus obtain f_{2N+1} in terms of f_{2N} and ϕ_{2N}.

EXERCISE

Obtain the corresponding recurrence relations when

$$f_N(c) = \min_u J(u).$$

9. Differential Equations

Let us consider the problem of minimizing the functional

$$J(x,y) = \int_0^T h(x,y)\,dt, \tag{1}$$

where x and y are connected by the differential equations

$$x' = g(x,y), \qquad x(0) = c. \tag{2}$$

Once again we suppose that the problem is properly formulated and set

$$\min_y J(x,y) = f(c,T). \tag{3}$$

Write

$$\int_0^T = \int_0^S + \int_S^T. \tag{4}$$

Supposing that S is small, we obtain the approximate functional equation

$$f(c,T) = \min_z [h(c,z)S + f(c + Sg(c,z), T - S)] + O(S^2), \tag{5}$$

where $z = y(0)$.

Passing to the limit as $S \to 0$, we obtain the nonlinear partial differential equation

$$f_T = \min_z [h(c,z) + (g(c,z), \operatorname{grad} f)], \tag{6}$$

with the initial condition $f(c,0) = 0$.

10. Quadratic Case

The results simplify greatly when $h(x, y)$ is quadratic in x and y and $g(x, y)$ is linear. Consider, for example, the minimization of

$$J(x) = \int_0^T [(x', x') + (x, Ax)]\, dt\,, \tag{1}$$

where A is positive definite and $x(0) = c$.
It is clear that

$$f(c, T) = \min_x J(x) = (c, R(T)\,c), \tag{2}$$

where $R(T)$ depends only on T. On the other hand, (9.6) yields

$$f_T = \min_z [(z, z) + (c, Ac) + (z, \operatorname{grad} f)]\,. \tag{3}$$

The minimization with respect to z is readily accomplished, yielding

$$z = -\operatorname{grad} f/2\,, \tag{4}$$

$$f_T = (c, Ac) - [(\operatorname{grad} f, \operatorname{grad} f)/4]\,. \tag{5}$$

Using (2) in (5), we obtain an ordinary differential equation, a Riccati equation,

$$R'(T) = A - R^2(T), \qquad R(0) = 0\,. \tag{6}$$

Using the value obtained above, we have

$$x'(0) = z = -R(T)\,c\,, \tag{7}$$

the desired initial condition.

11. Minimum Convolutions

Write

$$\min_u \int_S^T h(u, u'\ \ t)\, dt = f(a, b;\, S, T), \tag{1}$$

where u is subject to $u(S) = a$, $u(T) = b$. Then, as above,

$$f(a, b;\, S, T) = \min_c [f(a, c;\, S, R) + f(c, b;\, R, T)]\,, \qquad S \leqslant R \leqslant T. \tag{2}$$

If h is quadratic in u and u', then f is quadratic in a and b.

$$f(a,b;S,T) = a^2 r_{11}(S,T) + 2abr_{12}(S,T) + b^2 r_{22}(S,T). \tag{3}$$

Using (2), functional equations for the $r_{ij}(S,T)$ are obtained.

EXERCISE

Obtain these functional equations.

12. Tridiagonal Matrices

Let us consider the problem of solving the linear system

$$Ax = c, \tag{1}$$

where A is symmetric and tridiagonal, i.e.,

$$a_{ij} = 0, \qquad |i-j| > 1. \tag{2}$$

If we assume that A is positive definite, then the solution of (1) is equivalent to that of minimizing the quadratic form

$$Q(x) = (x, Ax) - 2(c, x). \tag{3}$$

Let us write α_i and β_i to denote

$$\alpha_i = a_{ii}, \qquad \beta_i = a_{i,i+1} = a_{i,i-1}. \tag{4}$$

Now we can write out (3) as

$$\begin{aligned} Q(x) = {} & \alpha_1 x_1{}^2 + 2\beta_1 x_1 x_2 + \alpha_2 x_2{}^2 + 2\beta_2 x_2 x_3 + \cdots \\ & + 2\beta_{n-1} x_{n-1} x_n + \alpha_n x_n{}^2 - 2c_1 x_1 - 2c_2 x_2 \cdots - 2c_n x_n. \end{aligned} \tag{5}$$

Let us define $Q_k(x, z)$ by

$$\begin{aligned} Q_k(x,z) = {} & \alpha_1 x_1{}^2 + 2\beta_1 x_1 x_2 + \cdots + 2\beta_{k-1} x_{k-1} x_k \\ & + \alpha_k x_k{}^2 - 2c_1 x_1 - 2c_2 x_2 \cdots - 2zx_k, \qquad k = 1, 2, \ldots,. \end{aligned} \tag{6}$$

Finally we define

$$f_k(z) = \min_{[x_1, x_2, \ldots, x_k]} Q_k(x, z). \tag{7}$$

Clearly since

$$Q(x) = Q_n(x, c_n), \tag{8}$$

$f_n(c_n)$ gives the minimum of $Q(x)$. Applying the techniques we have just presented, it is easy to verify that

$$f_k(z) = \min_{x_k} [\alpha_k x_k^2 - 2zx_k + f_{k-1}(c_{k-1} + \beta_{k-1} x_k)], \tag{9}$$

with

$$f_1(z) = z^2/\alpha_1. \tag{10}$$

Furthermore, we can establish that $f_k(z)$ is quadratic in z,

$$f_k(z) = r_k z^2 + 2s_k z + t_k, \tag{11}$$

which enables us to find recurrence equations for the coefficients r_k, s_k, and t_k. Much of what follows in the succeeding chapters will be based on matrix versions of these results.

EXERCISES

1. Prove (9).

2. Find the recurrence equations for the coefficients r_k, s_k and t_k. Use these results to obtain a recurrence equation for the x_k.

BIBLIOGRAPHY AND COMMENT

See

R. Bellman, *Introduction to the Mathematical Theory of Control Processes I: Linear Equations and Quadratic Criteria*, Academic Press, New York, 1967.

Section 7. See

R. Bellman, "Functional Equations in the Theory of Dynamic Programming XVII: Minimum Convolutions and Green's Functions," *J. Math. Anal. Appl.*, Vol. 33, 1971, pp. 497–499.

Section 8. The problem of computing 2^N with the least number of multiplications is an unsolved problem of mathematics. For some partial results, see

H. Kato, "On Addition Chains," Ph.D. Thesis, Univ. of Southern California, 1970.
D. E. Knuth, *The Art of Computer Programming II: Semi-Numerical Algorithms*, Addison-Wesley, Reading, Massachusetts, 1969.

Section 12. See

R. Bellman, *Introduction to Matrix Analysis*, 2nd ed., McGraw-Hill, New York, 1970.
R. S. Lehman, "Dynamic Programming and Gaussian Elimination," *J. Math. Anal. Appl.* Vol. 5, 1962, pp. 499–501.

Chapter 4

The Potential Equation

1. Introduction

In this chapter we wish to discuss some aspects of the potential equation,

$$u_{xx} + u_{yy} = 0, \qquad (x, y) \in R, \tag{1}$$

$$u = g(x, y), \qquad (x, y) \in \Gamma, \tag{2}$$

where Γ is the boundary of the region R in Fig. 1. We want to consider its connection with the minimization of the quadratic functional

$$D(u) = \int_R (u_x^{\,2} + u_y^{\,2})\, dR, \tag{3}$$

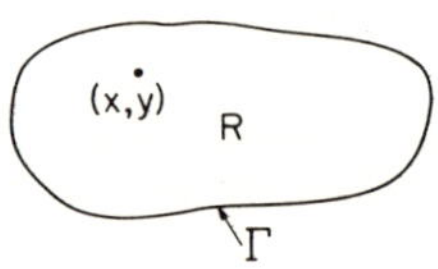

FIGURE 1

the Dirichlet functional, and a number of problems associated in this fashion.

This quadratic minimization problem can be approached directly and effectively in many cases by means of Rayleigh–Ritz procedures, thereby furnishing an indirect approach to the potential equation. We shall follow a different route based on the use of discretization, a route which has advantages in certain cases.

We will also be concerned with the Green's functions associated with the inhomogeneous equation

$$u_{xx} + u_{yy} = h(x, y)\,, \tag{4}$$

and the more general equation

$$u_{xx} + u_{yy} + k(x, y)\,u = h(x, y)\,. \tag{5}$$

2. The Euler–Lagrange Equation

We can readily show that the potential equation is the variational equation associated with $D(u)$. Let $v = 0$ on Γ so that the functions u and $u+v$ satisfy the same boundary conditions on Γ. Then

$$D(u + v) = D(u) + D(v) + 2\int_R (u_x v_x + u_y v_y)\, dR\,. \tag{1}$$

Applying Green's theorem, we see that the third term vanishes if u satisfies the potential equation. Hence if u satisfies (1.1) we have

$$D(u + v) > D(u) \tag{2}$$

for any nontrivial v. Thus, if (1.1) and (1.2) possess a solution, they possess a unique solution, since any solution minimizes $D(u)$.

Conversely, it can be shown from first principles that $D(u)$ possesses a minimum over the class of functions u such that $u_x, u_y \in L^2$ with u satisfying the specified boundary condition. Further it can be shown that the minimizing function is determined by (1.1) and (1.2).

Thus the problem of solving the potential equation and minimizing the Dirichlet functional $D(u)$ are equivalent. We shall focus henceforth on the minimization problem.

3. Inhomogeneous and Nonlinear Cases

It follows from the foregoing that the problem of solving the inhomogeneous problem

$$u_{xx} + u_{yy} = h(x,y), \qquad u(x,y) = 0, \quad (x,y) \in \Gamma, \tag{1}$$

and minimizing

$$D_1(y) = \int_R (u_x^2 + u_y^2 + 2h(x,y)u)\, dR, \tag{2}$$

subject to $u = 0$ on Γ, $u_x, u_y \in L^2(R)$, are equivalent.

Similarly, the problem of solving

$$u_{xx} + u_{yy} + k(x,y)u = 0, \tag{3}$$

and minimizing

$$D_2(u) = \int_R (u_x^2 + u_y^2 - k(x,y)u^2)\, dR, \tag{4}$$

are equivalent provided that $D_2(u)$ is a positive-definite functional. A sufficient condition for this is $\max_R |k(x,y)| < \lambda_1$, where λ_1 is the smallest characteristic value of the problem

$$u_{xx} + u_{yy} + \lambda_1 u = 0, \qquad u = 0, \quad (x,y) \in \Gamma.$$

Finally, the non linear equation

$$u_{xx} + u_{yy} - h(u) = 0, \tag{5}$$

may be associated with the functional

$$\int_R (u_x^2 + u_y^2 + 2g(u))\, dR, \tag{6}$$

where $g'(u) = h(u)$. We shall discuss this further in the chapter on quasilinearization.

4. Green's Function

The solution of the inhomogeneous problem

$$u_{xx} + u_{yy} = h(x,y), \qquad u = 0, \quad (x,y) \in \Gamma, \tag{1}$$

may be expressed in the form

$$u = \int_R k(x, y, x_1, y_1)\, h(x_1, y_1)\, dR. \tag{2}$$

The kernel k is called the *Green's function* associated with the particular equation and boundary condition.

For the foregoing case we have $k \leqslant 0$, a result we wish to use subsequently. Let us derive this fundamental property directly from the associated variational problem.

5. One-Dimensional Case

To illustrate the method of proof most simply, let us begin with the one-dimensional case. We wish to show that the solution of

$$u'' + g(t)\,u = h(t), \qquad u(0) = u(T) = 0, \tag{1}$$

is nonnegative in $[0, T]$ whenever $h(t)$ is negative, provided that the quadratic functional

$$J(u) = \int_0^T (u'^2 - g(t)\,u^2)\, dt \tag{2}$$

is positive definite for functions satisfying the boundary conditions above.

It follows from this hypothesis that (1) is the Euler equation associated with the minimization of

$$J_1(u) - \int_0^T [u'^2 - g(t)\,u^2 + 2h(t)\,u]\, dt. \tag{3}$$

The solution of (1) furnishes the absolute minimum of (3) over the class of functions such that $u' \in L^2(0, T)$, and the boundary conditions are satisfied.

Let us suppose that the stated result is not true, which means that u, a continuous function, is negative in some interval $[a, b]$; see Fig. 2. Here a may be 0 or b may be T.

Consider the new function v defined by

$$\begin{aligned} v &= u, & t &\notin [a, b], \\ v &= -u, & t &\in [a, b], \end{aligned} \tag{4}$$

The new function possibly has a discontinuous derivative at $t = a$ and $t = b$, but nevertheless is such that $v' \in L^2(0, T)$ and v satisfies the boundary conditions.

Write

$$J_1(u) = \int_0^T = \int_a^b + \int_S, \tag{5}$$

where $S = [0, T] - [a, b]$. The integral over S is clearly unchanged by replacing u by v. On the other hand, since $h(t) \leqslant 0$ we see that the integral over $[a, b]$ is decreased. Thus

$$J_1(v) < J_1(u). \tag{6}$$

This is a contradiction to the fact that u provides the *absolute* minimum of J_1. Hence $u \geqslant 0$ in $[0, T]$.

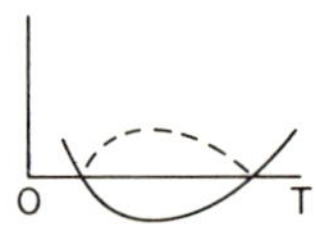

FIGURE 2

Since

$$u(t) = \int_0^T k(t, t_1) h(t_1)\, dt_1, \tag{7}$$

and $h(t)$ is an arbitrary negative function, this last fact implies that $k(t, t_1)$, the Green's function, is nonpositive for $0 \leqslant t,\ t_1 \leqslant T$, the desired result.

EXERCISES

1. Show that $k(t, t_1)$ is actually negative for $0 < t,\ t_1 < T$.

2. Show by similar argument that $k(t, t_1)$ has a variation-diminishing property. See

R. Bellman, "On Variation-Diminishing Properties of Green's Functions," *Boll. Un. Mat. Ital.*, Vol. 16, 1961, pp. 164–166.

3. Show that one is led to suspect these results by considering the minimization of $J_3(u) = \int_0^T [u'^2 + (u-h)^2]\, dt$. Why may this be interpreted as a *smoothing* operation?

4. Extend the foregoing argument to cover the equation

$$u'' - u - 2u^3 = h(t), \qquad u(0) = u(T) = 0.$$

5. What can we say about the solution of the equation

$$u'' - e^{-u} = h(t), \qquad u(0) = u(T) = 0,$$

if $h(t) \geqslant 0$?

6. Show that the same argument establishes a similar result for the difference equation

$$u_{n+1} - 2u_n + u_{n-1} + q_n u_n = h_n, \qquad n = 1, 2, \ldots, N-1, \quad u_0 = u_N = 0,$$

associated with the minimization of $\sum_{n=0}^{N-1} [(u_{n+1} - u_n)^2 - q_n u_n^2 + 2h_n u_n]$.

6. Two-Dimensional Case

The same method of proof can be used in higher dimensions. Consider the two-dimensional case. Let

$$D(u) = \int_R [u_x^2 + u_y^2 - q(x,y)u^2 + 2h(x,y)u] \, dR, \tag{1}$$

be the quadratic functional associated with the equation

$$u_{xx} + u_{yy} + q(x,y)u = h(x,y), \qquad u = 0, \quad (x,y) \in \Gamma. \tag{2}$$

We suppose that the quadratic part of $D(u)$ is positive definite.

Under this assumption we wish to show that $h \geqslant 0$ implies that $u \leqslant 0$. We proceed by contradiction. Let $u \geqslant 0$ in $R_1 \in R$ and consider the new function

$$\begin{aligned} v &= -u, \qquad p \in R_1, \\ v &= u, \qquad p \in R - R_1. \end{aligned} \tag{3}$$

As before, we see that $D(v) < D(u)$, a contradiction. Thus $h \geqslant 0$ implies $u \leqslant 0$, whence as before the Green's function is nonpositive.

EXERCISES

1. Show that the Green's function is actually negative for points inside R.

2. What is the corresponding variation-diminishing property?

3. Extend the result to suitable nonlinear equations as in the one-dimensional case.

4. What is the analogue of the result of Exercise 6 at the end of Section 5?

7. Discretization

One powerful approach to the study of the properties of differential equations is the use of associated difference equations. This procedure is particularly useful numerically when we contemplate the use of analog or digital computers. However, it is equally valuable for theoretical purposes. The fundamental observation is that the approximation

$$u_{xx}(x,y) \cong \frac{u(x+\Delta,y)+u(x-\Delta,y)-2u(x,y)}{\Delta^2},$$
$$u_{yy}(x,y) \cong \frac{u(x,y+\Delta)+u(x,y-\Delta)-2u(x,y)}{\Delta^2}, \tag{1}$$

converts the potential equation into a system of linear algebraic equations. This transformation, of course, is the beginning of the real problem, that of obtaining useful analytic and computational results from the linear system.

8. Rectangular Region

Let us see what this involves for a rectangular region (Fig. 3). Let integers M and N be chosen and the positive quantities Δ and δ be determined by $M\Delta = a$, $N\delta = b$. Let

$$u_{mn} = u(m\Delta, n\delta), \qquad m = 0,1,\ldots,M, \quad n = 0,1,\ldots,N. \tag{1}$$

The original boundary condition,

$$u(x,y) = g(x,y), \qquad (x,y) \in \Gamma, \tag{2}$$

yields the corresponding boundary conditions

$$\begin{aligned} u_{0n} &= g(0,n\delta), \qquad n = 0,1,\ldots,N, \\ u_{Mn} &= g(a,n\delta), \\ u_{m0} &= g(m\Delta,0), \qquad m = 0,1,\ldots,M, \\ u_{mN} &= g(m\Delta,b). \end{aligned} \tag{3}$$

We suppose that $g(x, y)$ is continuous so that $g(0, 0)$ and $g(m\Delta, N\delta)$ are unambiguously defined by the foregoing.

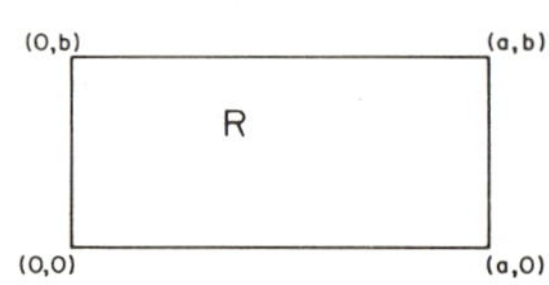

FIGURE 3

Upon setting $x = m\Delta$, $y = n\delta$, the potential equation yields the set of relations,

$$\frac{u((m+1)\Delta, n\delta) + u((m-1)\Delta, n\delta) - 2u(m\Delta, n\delta)}{\Delta^2} + \frac{u(m\Delta, (n+1)\delta) + u(m\Delta, (n-1)\delta) - 2u(m\Delta, n\delta)}{\delta^2} \cong 0. \quad (4)$$

Hence we obtain the linear difference equations

$$\frac{v_{m+1,n} + v_{m-1,n} - 2v_{mn}}{\Delta^2} + \frac{v_{m,n+1} + v_{m,n-1} - 2v_{mn}}{\delta^2} = 0,$$

$$m = 0, 1, \ldots, M, \qquad n = 0, 1, \ldots, N. \quad (5)$$

These relations, together with (3), constitute a system of linear algebraic equations for the quantities v_{mn}.

9. Rigorous Aspects

If Δ and δ are both sufficiently small, we suspect that

$$u(m\Delta, n\delta) \cong v_{mn}. \quad (1)$$

This can be established rigorously; references will be found at the end of the chapter.

EXERCISE

Show that, under the assumption that u possesses third derivatives, this is a stability result for the solution of (8.5).

10. Associated Minimization Problem

It is necessary to consider the existence and uniqueness of the solution of the linear system in (8.5). We can answer these questions readily by noting that the linear difference equations are the variational equations associated with the quadratic form

$$Q_{M,N}(v) = \sum_{m,n} \left[\left(\frac{v_{m+1,n} - v_{m,n}}{\Delta} \right)^2 + \left(\frac{v_{m,n+1} - v_{m,n}}{\delta} \right)^2 \right], \quad (1)$$

where m ranges over $0, 1, 2, \ldots, M$ and n over $0, 1, 2, \ldots, N$. The boundary values are determined as in (8.3). Since $Q_{M,N}(v)$ is clearly of positive-definite nature, it possesses a unique minimum value, attained by the unique solution of (8.5).

11. Approximation from Above

The foregoing minimization problem, in Section 9, is obtained from the original minimization problem, (associated with $D(u)$), by restricting attention to functions u for which u_x and u_y are constants over the rectangles

$$m\Delta \leqslant x \leqslant (m+1)\Delta, \qquad n\delta \leqslant y \leqslant (n+1)\delta. \tag{1}$$

It follows that the minimum over this subclass of functions is greater than or equal to the minimum over the original class of functions such that u_x and u_y are in $L^2(R)$. Hence, for any Δ, $\delta > 0$, we have

$$\min_v Q_{M,N}(v) \geqslant \min_u D(u). \tag{2}$$

12. Discussion

If we wish to obtain an accurate estimate for $u(m\Delta, n\delta)$ in the foregoing fashion, the quantities Δ and δ must be taken small. If we do this, however, we are left with a high order system of linear algebraic equations. The task of solving this system numerically is an onerous and unreliable one if straightforward techniques are employed.

Fortunately, the special structure of the system allows the use of some powerful and ingenious techniques in a number of cases. These methods, as all methods, possess advantages and disadvantages. We shall pursue a quite different approach here.

13. Semidiscretization

It is occasionally advantageous to introduce discretization in one variable only. Thus, for example, we may wish to use a grid only in the y-direction. The potential equation is then replaced by

$$w_{xx}^{(m)} = \frac{w^{(m-1)} + w^{(m+1)} - 2w^{(m)}}{\Delta^2}, \tag{1}$$

where $w^{(m)} = w(x, m\Delta)$.

For the region R_1, where $0 \leqslant x \leqslant a$, $0 \leqslant y \leqslant b$, the integer m ranges from 1 to $M-1$ with the two-point boundary conditions

$$w^{(m)}(0) = a_m, \quad w^{(m)}(a) = b_m, \quad w^{(0)}(x) = g(x), \quad w^{(M)}(x) = h(x),$$
$$m = 0, 1, \ldots, M, \quad 0 \leqslant x \leqslant a. \tag{2}$$

The quantities a_m, b_m, $g(x)$ and $h(x)$ are immediately determined from (8.2).

The solution of linear differential equations of this nature can be effected by means of the solution of M-dimensional linear algebraic systems. The associated variational problem is the minimization of

$$\int_0^a \left[\sum_{m=1}^{M-1} (w_x^{(m)})^2 + \sum_{m=1}^{M-1} \left(\frac{w^{(m+1)} - w^{(m-1)}}{\Delta} \right)^2 \right] dx \,. \tag{3}$$

The positive-definite nature of this functional ensures the existence and uniqueness of the solution of (1) and (2).

14. Irregular Grid

There is no necessity, of course, to employ a regular grid. Let y assume the values

$$y = y_0, y_1, \ldots, y_{n-1}, y_n, \tag{1}$$

where $0 = y_0 < y_1 < y_2 < \cdots < y_{n-1} < y_n = a$, and consider a "quadrature" formula

$$u_{yy} \cong \sum_{j=0}^{n} w_j u(x, y_j), \qquad y = y_0, y_1, \ldots, y_n. \tag{2}$$

One way to determine the quantities w_i and y_i is to require that the formula be exact for polynomials of degree n or less. With the aid of this approximation the potential equation becomes

$$(z_i)_{xx} - \sum_j w_{ij} z_j = 0. \tag{3}$$

Once again we face a two-point boundary value problem. The advantage, however, lies in the fact that the order of the differential equation can be taken to be far less than the order of that in Section 13.

Spline approximations can also be used.

EXERCISE

Is there an associated variational problem?

15. Solution of the Difference Equations

The linear difference (or finite difference) equations, (8.5), can be written in matrix–vector form

$$Av = b, \tag{1}$$

where v is the $(M-1)(N-1)$-dimensional vector of the unknown $\{v_{ij}\}$ at the interior points

$$v = \begin{bmatrix} v_{1,1} \\ \vdots \\ v_{1,N-1} \\ v_{2,1} \\ \vdots \\ v_{M-1,N-1} \end{bmatrix}. \tag{2}$$

The matrix A is determined by the finite difference approximation (8.4), and the grid spacing while the vector b is determined by the boundary conditions. The structure of A has many special properties which we will investigate in subsequent chapters. At the moment we again note that A is nonsingular which allows us to write v as

$$v = A^{-1}b. \tag{3}$$

Comparing this result with (4.2) it is clear that A^{-1} is the discrete analog of the Green's function. All the elements of A^{-1} are positive and thus this discrete Green's function possesses a property analogous to the negativity of $k(t, t_1)$. The proof follows the lines of the proof for the continuous case.

Since the dimension of A is large when an accurate numerical result is desired, a solution of (1) by conventional direct methods, such as Gaussian elimination, is impractical. Subsequently we will employ the block tridiagonal structure of A to our advantage. Going back to (8.5), we see that a row of A can have at most five nonzero entries. Due to this sparseness, iterative methods can be effectively employed to solve (1) numerically.

16. Iterative Solutions

A linear iterative method for the solution of the linear system $v = Dv+f$ can be described in the form

$$v^{(k+1)} = Dv^{(k)} + f, \tag{1}$$

where $v^{(0)}$ is some initial guess. If $\{v^{(k)}\}$ converges, it converges to the solution of the original linear equation

$$[I - D]v = f. \tag{2}$$

Thus if there exists a nonsingular matrix T such that

$$T[I - D] = A, \qquad Tf = b, \tag{3}$$

when the sequence $v^{(k)}$ converges we will have a solution of (15.1). Since each iteration of (1) requires one multiplication by D, we restrict ourselves in practice to those matrices A and T for which D is sparse.

In order to investigate under what circumstances (1) converges, let us define an error

$$e^{(k)} = v - v^{(k)}. \tag{4}$$

Then using (1) and (2)

$$\begin{aligned} e^{(k+1)} &= v - v^{(k+1)} \\ &= v - Dv^{(k)} - f \\ &= v - Dv^{(k)} - [I - D]v, \end{aligned} \tag{5}$$

whence

$$e^{(k+1)} = De^{(k)}, \tag{6}$$

and

$$e^{(k)} = D^k e^{(0)}. \tag{7}$$

Thus (1) converges for all $e^{(0)}$ if and only if all the characteristic values of D are of absolute value less than unity. Let λ_k be the kth characteristic value of D. We define the spectral radius of D, $\rho(D)$, to be

$$\rho(D) = \max_k |\lambda_k|. \tag{8}$$

Then (1) converges if and only if

$$\rho(D) < 1. \tag{9}$$

EXERCISES

1. Let A and b be defined by (8.5). Show that if we let $T = I$ the resulting matrix D satisfies (3). This defines the Jacobi method.

2. Show that each iteration of the Jacobi method requires $O(NM)$ multiplications.

17. Limitations of the Iterative Approach

We have yet to discuss how the matrix D and the vector f are determined from A and b. Although our emphasis will be on direct methods, we will discuss some of the various iterative methods in Chapter 6. However, at this point we can recognize some of the limitations of the iterative methods.

The matrix D for the simple methods such as the Jacobi or Gauss–Seidel methods is determined directly from the matrix A. Thus the convergence of these methods depends only on the original partial differential equation and the discretization. For a large class of equations, including the potential equation, these simple methods can be shown to converge for any step size. Unfortunately, however, these methods may converge very slowly as the step size is reduced, thus effectively limiting the accuracy of the results.

The more sophisticated iterative methods such as successive over-relaxation, or the alternating implicit methods, require the determination of one or more parameters chosen to make the spectral radius of D as small as possible. For equations other than the potential equation over a rectangular region the determination of these parameters is a difficult analytic and computational task. In terms of the actual cost of solving a given problem it may not be worth expending the necessary effort to determine these parameters precisely. However, if we are in the neighborhood of an optimum, determination of the optimal parameters is not very important and good guesses may often be obtained as a by-product of early iterations. Furthermore, methods such as the alternating direction implicit methods may even fail to converge if the region of interest is of any more general type than a rectangle.

When we discuss quasilinearization and identification, we will encounter some other reasons why a direct method is preferable to an iterative one. In general, however, there is a basis for prefering a direct method solely on the grounds that a direct method is easy to understand, easy to program, and is usually operational. At a time when computers are operating in microseconds and high speed memories are in the hundreds of thousands of words, the ease with which numerical methods can be used by the scientist and engineer, and not just the numerical analyst, is a primary consideration.

MISCELLANEOUS EXERCISES

1. Let $L(u)$ be a linear functional, continuous in u for $u \in L^2(0,1)$. Consider the minimization of the quadratic functional $\int_0^1 u^2\,dt + L(u)$ and thereby establish the Riesz representation theorem, $L(u) = \int_0^1 uv\,dt$, for suitable $v \in L^2(0,1)$.

2. Consider the equation $u''=0$, $u(0)=a$, $u(T)=b$ and the associated difference equation $u_{n+1}+u_{n-1}-2u_n=0$, $n=1,2,\ldots,N-1$, $u_0=a$, $u_N=b$. Consider the method of successive approximations

$$u_n^{(k)} = \frac{u_{n+1}^{(k-1)}+u_{n-1}^{(k-1)}}{2}, \qquad u_0^{(k)} = a, \qquad u_N^{(k)} = b.$$

Does $u_n^{(k)}$ converge as $k\to\infty$? Monotonically?

3. Consider the equations of (8.4) for $\Delta=\delta$. Show that they may be put in the form

$$v_{m,n} = \tfrac{1}{4}(v_{m+1,n} + v_{m-1,n} + v_{m,n+1} + v_{m,n-1}).$$

4. Consider the method of successive approximations

$$v_{m,n}^{(k)} = \tfrac{1}{4}(v_{m+1,n}^{(k-1)} + \cdots).$$

Does $v_{m,n}^{(k)}$ converge as $k\to\infty$? Monotonically?

BIBLIOGRAPHY AND COMMENT

Section 1. There are a number of books which discuss the Dirichlet problem and the calculus of variations. See, for example,

R. Courant, *Dirichlet's Principle*, Wiley (Interscience), New York, 1950.

R. Courant and D. Hilbert, *Methods of Mathematical Physics*, Wiley (Interscience), New York, 1953.

N. I. Akhieser, *The Calculus of Variations*, Ginn (Blaisdell), Boston, Massachusetts, 1962.

I. M. Gelfand and S. V. Fomin, *Calculus of Variations*, Prentice-Hall, Englewood Cliffs, New Jersey, 1963.

For alternate approaches see

R. Bellman, *Methods of Nonlinear Analysis*, Academic Press, New York, 1970.

R. Varga, "Accurate Numerical Methods for Nonlinear Boundary Value Problems," *SIAM–AMS Proc.*, Vol. 2, 1970.

Sections 4–6. See

R. Bellman, *Introduction to the Mathematical Theory Control Processes, I: Linear Equations and Quadratic Criteria*, Academic Press, New York, 1967.

Section 8. See

J. H. Bramble, B. E. Hubbard, and Vidar Thomée, *Math. Comp.*, Vol. 23, 1969, pp. 695–710.

Additional references are given there.

Section 9. For proofs of convergence see

E. Isaacson and H. B. Keller, *Analysis of Numerical Methods*, Wiley, New York, 1966.

Additional references are contained in the chapter by D. Young in

J. Todd, ed., *Survey of Numerical Analysis*, McGraw-Hill, New York, 1962.

See also

R. D. Richtmyer and K. W. Morton, *Difference Methods for Initial Value Problems*, 2nd ed., Wiley (Interscience), New York, 1967.
L. Collatz, *The Numerical Treatment of Differential Equations*, 3rd ed., Springer-Verlag, Berlin and New York, 1960.
L. Collatz, *Functional Analysis and Numerical Mathematics*, Academic Press, New York, 1966.

Section 10. This is an example of approximation in policy space.

Section 13. The semidiscretization approach is also known as the *method of lines*. This approach is very popular in the Russian literature. The reader should be aware that virtually all the methods which follow could also be posed in terms of the method of lines rather than via the finite difference approach. See

L. V. Kantorovich and V. I. Krylov, *Approximate Methods of Higher Analysis*, Wiley (Interscience), New York, 1958.

Section 14. See

R. Bellman, and J. Casti, "Differential Quadrature and Long Term Integration," *J. Math. Anal. Appl.*, Vol. 34, 1971, pp. 235–238.

Section 16. This approach is taken by R. Varga to analyze a number of standard techniques. See

R. Varga, *Matrix Iterative Analysis*, Prentice-Hall, Englewood Cliffs, New Jersey, 1962.
A. Ralston and H. S. Wilf, *Mathematical Methods for Digital Computers*, Vols. 1 and 2, Wiley, New York, 1965 and 1967.

Section 17. John Todd points out that ADI only works if certain matrices commute and we are in standard academic situations. In practice, however, the situation is often favorable.

Chapter 5

Dynamic Programming and Elliptic Equations

1. The Potential Equation

In this chapter, we will present an approach to the numerical solution of some types of elliptic equations based on the theory of dynamic programming. We start with the problem introduced in Chapter 4, the potential equation

$$\nabla^2 u = u_{xx} + u_{yy} = 0\,, \tag{1}$$

over a region R with the boundary conditions

$$u(x, y) = f(x, y)\,, \tag{2}$$

on Γ the boundary of R.

We have seen that (1) is the Euler equation associated with the minimization of the quadratic functional

$$D(u) = \iint_R [u_x^2 + u_y^2]\, dR\,, \tag{3}$$

subject to (2).

For initial purposes of exposition we will assume that R is the rectangle $0 \leqslant x \leqslant a$, $0 \leqslant y \leqslant b$.

Subsequently, we will treat some irregular regions as combinations of rectangular regions.

2. Discretization

Our method will be to determine the minimum of a discretized version of the Dirichlet functional corresponding to the original potential equation. We will assume for convenience that a and b are such that

$$a = Nh\,, \qquad b = Mh\,, \tag{1}$$

where M and N are integers. The modifications necessary if this is not the case are simple and do not effect the subsequent development. We seek a function at the points of a finite grid (or mesh). Specifically, we are interested in the points

$$u_{ij} = u(ih,jh)\,, \qquad i = 0, 1, \ldots, N, \quad j = 0, 1, \ldots, M\,. \tag{2}$$

Since we want a solution defined only at these $(N+1)(M+1)$ points, we would like to replace the partial derivatives at these points by terms involving only the function values at these points. Using a Taylor series expansion we have

$$\frac{\partial u_{ij}}{\partial x} = \frac{u_{ij} - u_{i-1,j}}{h} + O(h), \qquad \frac{\partial u_{ij}}{\partial y} = \frac{u_{ij} - u_{i,j-1}}{h} + O(h), \tag{3}$$

a standard finite difference approximation.

Thus, to obtain a discrete version of the functional in (1.3) we replace the partial derivatives appearing by the expressions in (3) and assume that the derivatives are constant between grid points. Equivalently, we could apply rectangular integration to (1.3) and then use (3). Either way the double integral in the Dirichlet functional can be approximated by the double sum

$$J(u) = \sum_{i=1}^{N} \sum_{j=1}^{M} [(u_{ij} - u_{i,j-1})^2 + (u_{ij} - u_{i-1,j})^2]\,. \tag{4}$$

We can easily verify that the discretization error is $O(h^2)$. The discretized functional is subject to the original boundary conditions and thus the values u_{0j}, u_{i0}, u_{Nj}, and u_{iM} are determined by (1.2).

We are left with the finite dimensional problem of finding

$$\min_{\{u_{ij}\}} J(u)\,, \qquad i = 1, 2, \ldots, N-1\,, \quad j = 1, 2, \ldots, M-1\,. \tag{5}$$

It should be clear that since we are interested only in the minimizing $\{u_{ij}\}$ and not the minimum value of $J(u)$ itself we can remove constant terms from (4) without changing the set of minimizing $\{u_{ij}\}$. It will turn out to be convenient to remove from (4) all terms of the form $(u_{iM} - u_{i-1,M})^2$ which are fixed by the boundary conditions. In this manner we replace (4) by the more convenient functional

$$J(u) = \sum_{i=1}^{N} \left[\sum_{j=1}^{M} (u_{ij} - u_{i,j-1})^2 + \sum_{j=1}^{M-1} (u_{ij} - u_{i-1,j})^2 \right]. \qquad (6)$$

3. Matrix–Vector Formulation

Although we have now completely discretized the original problem, it is still not in a form which can easily be solved. We will first put the problem into matrix–vector form and then convert it into a multistage decision process which can be handled by dynamic programming.

Since the functional is a quadratic form, we can conveniently rewrite the problem using inner product notation. First, we define the vectors of interior values

$$u_R = \begin{bmatrix} u_{R1} \\ u_{R2} \\ \vdots \\ u_{R,M-1} \end{bmatrix}, \qquad R = 1, 2, \ldots, N-1, \qquad (1)$$

and note that the vectors u_0 and u_N are given by the boundary conditions. Now we define a symmetric matrix of order $M-1$, Q; a set of $(M-1)$-dimensional vectors, r_R, and scalars, s_R, as follows

$$Q = (q_{ij}) \qquad \text{where} \quad q_{ij} = \begin{cases} 2, & i = j, \\ -1, & |i-j| = 1, \\ 0, & \text{otherwise}, \end{cases}$$

$$r_R = [r_{Rj}] \qquad \text{where} \quad r_{Rj} = \begin{cases} u_{R0}, & j = 1, \\ u_{RM}, & j = M-1, \\ 0, & \text{otherwise}, \end{cases} \qquad (2)$$

$$s_R = u_{R0}^2 + u_{RM}^2 .$$

Clearly, Q is a fixed matrix, while r_R and s_R are functions of only the boundary conditions. With this notation, $J(u)$ becomes, in inner product form,

$$J(u) = \sum_{R=1}^{N} [(Qu_R, u_R) - (2r_R, u_R) + s_R + (u_R - u_{R-1}, u_R - u_{R-1})]. \qquad (3)$$

The minimization is now carried out over the *vectors* u_R $(R = 1, 2, \ldots, N-1)$. The original boundary conditions are reduced to two point conditions.

4. Dynamic Programming

Instead of the original problem, let us consider the sequence of variational problems

$$f_R(v) = \min_{[u_R, u_{R+1}, \ldots, u_{N-1}]} \sum_{i=R}^{N} [(Qu_i, u_i) - (2r_i, u_i) + s_i$$
$$+ (u_i - u_{i-1}, u_i - u_{i-1})], \qquad R = 1, 2, \ldots, N-1, \tag{1}$$

where v is defined as

$$v = u_{R-1}. \tag{2}$$

Clearly, by comparison with (3.3) we have

$$f_1(u_0) = \min_{\{u_R\}} J(u). \tag{3}$$

Each of the problems in (1) can be regarded as the solution of the potential equation over a truncated rectangle with the left boundary condition, v, unspecified as in Fig. 1. This is our fundamental imbedding.

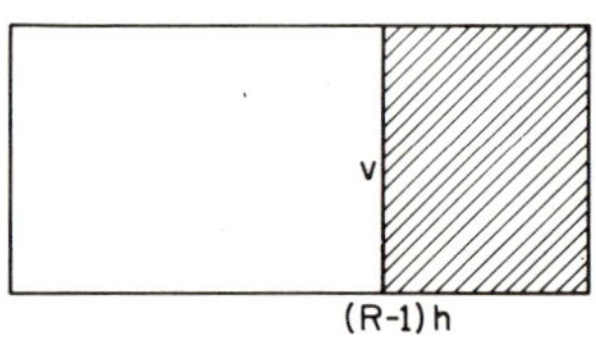

FIGURE 1

Let us next find a relationship between the function $f_R(v)$ and $f_{R+1}(v)$. Noting that only the first four terms of (1) depend on u_R, we can write

$$f_R(v) = \min_{[u_R, u_{R+1}, \ldots, u_{N-1}]} \Big[(Qu_R, u_R) - 2(r_R, u_R) + s_R + (u_R - v, u_R - v)$$
$$+ \sum_{i=R+1}^{N} \{(Qu_i, u_i) - 2(r_i, u_i) + s_i + (u_i - u_{i-1}, u_i - u_{i-1})\}\Big]. \tag{4}$$

The minimization over $u_{R+1}, u_{R+2}, \ldots, u_{N-1}$ can be moved past the first four terms, resulting in the expression

$$f_R(v) = \min_{u_R} \Big\{ (Qu_R, u_R) - 2(r_R, u_R) + s_R + (u_R - v, u_R - v) + \min_{u_{R+1}, u_{R+2}, \ldots, u_{N-1}} \Big[\sum_{i=R+1}^{N} (Qu_i, u_i) - 2(r_i, u_i) + s_i + (u_i - u_{i-1}, u_i - u_{i-1}) \Big] \Big\}. \tag{5}$$

However, by comparing the terms to be minimized over $u_{R+1}, u_{R+2}, \ldots, u_{N-1}$ with the definition in (1) of $f_R(v)$ we see that (5) can be rewritten as

$$f_R(v) = \min_{u_R} [(Qu_R, u_R) - 2(r_R, u_R) + s_R + (u_R - v, u_R - v) + f_{R+1}(u_R)]. \tag{6}$$

This result may also be considered to be a consequence of the principle of optimality. The choice of u_R must balance the cost of the present stage, the first four terms of (5), against the cost of continuing the process starting with u_R, and proceeding through $N-R$ additional stages.

Since u_N is fixed by the boundary conditions, we have

$$f_N(v) = (Qu_N, u_N) - 2(r_N, u_N) + s_N + (u_N - v, u_N - v). \tag{7}$$

Theoretically, we could now solve for the minimizing $\{u_R\}$ by starting with (7) and solving (6) backwards until $f_1(u_0)$ is determined. However, we have yet to exploit fully the quadratic nature of the variational problem. When we do so, the analytic and computational features will simplify considerably.

5. Recurrence Equations

It can easily be verified inductively that $f_R(v)$ must be quadratic in v,

$$f_R(v) = (A_R v, v) - 2(b_R, v) + c_R, \tag{1}$$

where A_R is a symmetric matrix. Furthermore the quantities A_R, b_R, and c_R are independent of v. Substituting this quadratic form for $f_{R+1}(u_R)$ in (4.5) we have

$$f_R(v) = \min_{u_R} [(Qu_R, u_R) - 2(r_R, u_R) + s_R + (u_R - v, u_R - v) + (A_{R+1} u_R, u_R) - 2(b_{R+1}, u_R) + c_R]. \tag{2}$$

The expression can be readily differentiated in order to find the minimizing u_R. The fact that this u_R actually yields a minimum can be easily verified from the positive definiteness of the functional. This is a point which will be more fully discussed later. The minimizing u_R is found to be

$$u_R = [Q + A_{R+1} + I]^{-1}(v + b_{R+1} + r_R). \tag{3}$$

Upon using this expression for u_R in (2) we obtain

$$\begin{aligned} f_R(v) = {} & ([I - [I + Q + A_{R+1}]^{-1}]v, v) \\ & - (2[I + Q + A_{R+1}]^{-1}(b_{R+1} + r_R), v) + c_{R+1} + s_{R+1} \\ & - ([I + Q + A_{R+1}]^{-1}(b_{R+1} + r_R), b_{R+1} + r_R). \end{aligned} \tag{4}$$

Thus by comparison with (1), we obtain the relations

$$\begin{aligned} A_R &= I - [I + Q + A_{R+1}]^{-1}, \\ b_R &= [I + Q + A_{R+1}]^{-1}(b_{R+1} + r_R), \\ &= [I - A_R](b_{R+1} + r_R), \\ c_R &= c_{R+1} + s_R - ([I + Q + A_{R+1}]^{-1}(b_{R+1} + r_R), b_{R+1} + r_R) \\ &= c_{R+1} + s_R - (b_R, b_{R+1} + r_R). \end{aligned} \tag{5}$$

From (4.7) we get the initial conditions

$$A_N = I, \qquad b_N = u_N, \qquad c_N = ([I + Q]u_N, u_N) - 2(r_N, u_N) + s_N. \tag{6}$$

Since v is by definition u_{R-1}, (3) becomes

$$u_R = [I + Q + A_{R+1}]^{-1}(u_{R-1} + b_{R+1} + r_R), \tag{7}$$

or

$$u_R = [I - A_R]u_{R-1} + b_R. \tag{8}$$

Since we are primarily interested in u_R, we can ignore the equation for c_R.

6. The Calculations

The calculations proceed as follows. Equations (5.5) are solved successively starting with (5.6) until A_1 and b_1 have been determined. All intermediate values of A_R and b_R are stored. Relation (5.8) is an initial value problem starting with u_0, the second boundary condition, A_1 and b_1. This equation is iterated using the last computed value u_R and the stored values of A_R and b_R. A sample computer program can be found in the appendix. As is typical of the dynamic programming approach, a two-point boundary value problem has been converted to two *initial value* problems. We will now show that the method always provides a solution and is computationally stable.

7. Nonsingularity

Let us prove that all the matrices to be inverted in the course of the foregoing process are nonsingular. The proof will be based on the fact that the matrices $[I+Q+A_{R+1}]$ are positive definite and of spectral radius less than one.

We will use the notation

$$A > B \tag{1}$$

to mean that $A-B$ is positive definite. We will first show that Q is positive definite. Let x be an arbitrary nonzero vector. Then, using the inner product

$$\begin{aligned}(Qx, x) &= \sum_{i=1}^{N}\sum_{j=1}^{N} q_{ij}x_i x_j \\ &= 2\sum_{i=1}^{N} x_i^2 - 2\sum_{i=2}^{N} x_i x_{i-1} \\ &= \sum_{i=1}^{N} x_i^2 + \sum_{i=2}^{N}(x_i - x_{i-1})^2 + x_1^2 + x_N^2 \\ &> 0,\end{aligned} \tag{2}$$

whence Q is positive definite.

We will proceed with the proof inductively. We have

$$A_{N-1} = I - [I+Q]^{-1}. \tag{3}$$

Since

$$Q > 0, \tag{4}$$

it follows that

$$I + Q > I. \tag{5}$$

Thus, $[I+Q]$ is nonsingular and A_{N-1} exists. Since $[I+Q]$ is positive definite, we have

$$[I+Q]^{-1} > 0, \tag{6}$$

and from (5)

$$[I+Q]^{-1} < I. \tag{7}$$

Thus

$$0 < A_{N-1} < I. \tag{8}$$

Assume now that

$$0 < A_R < I, \tag{9}$$

for some $R < N-1$. Then

$$[I + Q + A_R] > I, \tag{10}$$

and therefore

$$0 < [I + Q + A_R]^{-1} < I. \tag{11}$$

Since

$$A_{R-1} = I - [I + Q + A_R]^{-1}, \tag{12}$$

we have

$$0 < A_{R-1} < I, \tag{13}$$

which completes the induction.

Thus, all the matrices we must invert are positive definite and therefore nonsingular.

EXERCISE

Show that $(X+Y)^{-1} \cong X^{-1} - X^{-1}YX^{-1}$ if Y is small.

8. Stability

We will take *stability* to mean that the effect of an error made in one stage of the computation is not propagated into larger errors in latter stages of the computation. In other words, local errors are not magnified by further computation.

Let us first examine the matrix recurrence equation

$$A_R = I - [I + Q + A_{R+1}]^{-1}, \tag{1}$$

and let us assume that a small error has been made. It follows that we are actually employing the recurrence relation

$$\tilde{A}_R = I - [I + Q + \tilde{A}_{R+1}]^{-1}, \tag{2}$$

where

$$\tilde{A}_R = A_R + E_R, \tag{3}$$

is the desired solution plus an error term E_R. Using (1)–(3) we find the error at the next stage is given by

$$E_R = [I + Q + A_{R+1}]^{-1} - [I + Q + A_{R+1} + E_{R+1}]^{-1}. \tag{4}$$

After some manipulation, this expression becomes

$$E_R = [I + Q + \tilde{A}_{R+1}]^{-1} E_{R+1} [I + Q + A_{R+1}]^{-1}. \tag{5}$$

Under the assumptions that the initial error is sufficiently small and that the matrices are not ill-conditioned, we have by (5.5)

$$E_R \cong [I - A_{R+1}] E_{R+1} [I - A_{R+1}]. \tag{6}$$

Taking norms in (6) we can write

$$\| E_R \| \leqslant \| I - A_{R+1} \|^2 \| E_{R+1} \|. \tag{7}$$

Since all the matrices are symmetric (as a consequence of theory and the choice of numerical procedures) we can use the norm

$$\| M \| = \rho^{1/2}(M^2), \tag{8}$$

where $\rho(M^2)$ denotes the modulus of M^2. Then (7) becomes

$$\| E_R \| \leqslant \rho([I - A_R]^2) \| E_{R+1} \|. \tag{9}$$

However, from Section 6 we know

$$0 < I - A_R < I, \tag{10}$$

and thus

$$\rho([I - A_R]^2) < 1. \tag{11}$$

Finally

$$\| E_R \| < \| E_{R+1} \|, \tag{12}$$

from which we can conclude that the matrix equation is stable.

Using the same approach we can analyze the vector recurrence relation

$$b_R = [I - A_R] (b_{R+1} + r_R). \tag{13}$$

Again, if an error has been made we are actually employing the recurrence relation

$$\tilde{b}_R = [I - A_R] (\tilde{b}_{R+1} + r_R), \tag{14}$$

with

$$\tilde{b}_R = b_R + e_R. \tag{15}$$

In accordance with what we did above we can assume that the value of $I-A_R$ in (14) is essentially exact. Carrying out the same operations as before we find

$$e_R = [I - A_R]\, e_{R+1}, \tag{16}$$

and upon taking norms

$$\| e_R \| < \| e_{R+1} \| . \tag{17}$$

Thus this equation is also stable.

Finally, we carry out the same analysis on the final equation

$$u_R = [I - A_R]\, u_{R-1} + b_R . \tag{18}$$

If e_R now denotes the error in u_R we see

$$e_R = [I - A_R]\, e_{R+1} \tag{19}$$

and once again

$$\| e_R \| < \| e_{R+1} \| . \tag{20}$$

We conclude that the method is stable.

EXERCISES

1. Under what condition is the solution of the matrix vector differential equation

$$x' = Ax + b$$

stable?

2. Discuss the stability of the matrix Riccati equation

$$R' = Q - R^2, \qquad R(0) = 0,$$

when Q is positive definite.

9. Discussion

This stability is not unexpected. In fact, under some very general conditions, it is essentially guaranteed by the theory of dynamic programming. We can argue this point as follows. The solution we seek represents an optimal path associated with the functional from which it arose. The principle of optimality tells us how to proceed on an optimal path starting at an arbitrary state. If, due to an error, we find ourselves in some state other than the one we computed, we will thus automatically proceed in an optimal way with regard to this new state. Hence, at every stage of the procedure, we perform an optimization starting with the actual state. This multistage decision procedure tends to prevent error propagation.

In addition to the obvious fact that the stability of the method allows us to compute solutions of our equations effectively, there is further advantage. Since all the errors are essentially local errors, for a rough step size, h, the error will consist mostly of local truncation. Since the truncation is known to be $O(h^2)$, a "deferred approach to the limit" technique can be expected to improve the results significantly. This technique will be discussed in Section 12.

It can easily be shown that the dynamic programming method solves the same set of linear algebraic equations as given by the standard finite difference approximation of the potential equation. This point will become obvious in Chapter 6.

10. Efficiency

It is fairly easy to count the number of operations required by the foregoing method. We will consider only the necessary multiplications and divisions. The matrices to be inverted are not sparse. Hence, each inversion of a matrix of order M will take approximately $M^3/2$ multiplications and divisions. Thus, a total of $NM^3/2$ multiplications and divisions are necessary to solve the matrix recurrence equations, neglecting lower order terms. The two vector equation can be seen to require a total of $2NM^2$ multiplications and divisions.*

This number of operations at first sight seems to be greater than the number of operations required by either successive overrelaxation (SOR) or alternating direction implicit (ADI) techniques, namely $14N^3 \log N$ and $40N^2 \log^2 N$ operations respectively for $M = N$. However, these numbers are a bit deceptive. Both the SOR and ADI techniques are *iterative* and require suitable choices of certain parameters to ensure convergence. If these parameters are not chosen almost optimally, the methods may not converge or may require many additional iterations. The optimal choice

* We can significantly reduce the number of computations [to $O(NM^2)$] if we note that the matrix which diagonalizes Q also diagonalizes each A_R and thus all computations can be carried out in terms of the characteristic values of Q. We will not consider this technique, important as it is, since it is restricted to constant coefficient equations. In Chapter 8, we will consider special computational methods of this type.

of parameters is in itself a difficult mathematical problem since these parameters are functions of the equation, the boundary conditions, and the regions. In Chapter 11, when we discuss nonlinear equations, we will see the importance of this observation.

Perhaps more important is what we mean by *efficiency*. It is often the case that we are required to solve the same equation with many different sets of boundary conditions, and perhaps changes in equation parameters. Let us for now consider what happens when we try to solve the potential equation a second time with different boundary conditions. The matrix recurrence equation which requires most of the computation time is *independent* of the boundary conditions. Thus, there is no need to recompute its solution and the entire solution procedure in each particular case is reduced to resolving the two vector equations. Only $2NM^2$ multiplications are required to solve this part of the problem. This theme will recur frequently in all that follows.

Finally let us comment on the storage difficulties. We are forced to store the values of the matrices A_R. Even though they are symmetric, this nonetheless requires a great deal of storage, an example of the "curse of dimensionality." However, since the matrices are each recalled only once and sequentially, slow access storage can be used very effectively. Furthermore, let us note that in modern multiprocessing machines, the use of slow access storage is essentially free.

11. Example

As a simple example of the foregoing method the potential equation was solved on the unit square with the boundary conditions

$$\begin{aligned} u(x,0) &= 1, \qquad & u(0,y) &= 0, \\ u(x,1) &= 0, \qquad & u(1,y) &= 0. \end{aligned} \tag{1}$$

The required matrix inversions were carried out using the Gauss–Jordan elimination procedure which was available in the system library. The matrices A_R and vectors b_R were stored on a high speed disc.

Some typical results are shown in Table I for the dynamic programming method as compared with successive overrelaxation (27 iterations) and the series solution (30 terms). Both the dynamic programming and successive overrelaxation methods used a step size of $\frac{1}{16}$. All three computations required computing times of the same order of magnitude. However in view of the next section and the special methods we shall discuss in Chapter 8, it would be inappropriate to discuss computing times here.

TABLE I

Point	Dynamic programming	Successive overrelaxation	Series
$(\frac{1}{4}, \frac{1}{4})$	0.43178	0.43178	0.43203
$(\frac{1}{2}, \frac{1}{4})$	0.53932	0.53933	0.54053
$(\frac{1}{2}, \frac{3}{4})$	0.09564	0.09564	0.09541
$(\frac{3}{4}, \frac{1}{2})$	0.18251	0.18253	0.18203

12. Deferred Passage to the Limit

We pointed out in Section 8 that essentially all of the errors in the final result are local errors. If we do not choose h too small, something to be avoided on grounds of storage and time, then most of the error is local truncation.

Let us assume that the solution of the discretized problem, $u(x, y, h)$, depends in an analytic fashion on h^2, and see what use we can make of this,

$$u(x,y,h) = u(x,y,0) + u_1(x,y)h^2 + u_2(x,y)h^4 + \cdots. \tag{1}$$

The quantity of interest is $u(x, y, 0)$. One estimate of this is naturally

$$u(x,y,0) \cong u(x,y,h), \tag{2}$$

with an error that is $O(h^2)$. One way to improve this estimate is to use a smaller value of h. Since, however, the computational time is proportional to $1/h^4$, there are obvious disadvantages to this direct procedure.

Let us use (1) to obtain a better estimate than that appearing in (2). Write

$$u(x,y,h/2) = u(x,y,0) + u_1(x,y)(h^2/4) + u_2(x,y)(h^4/16) + \cdots. \tag{3}$$

Then

$$u(x,y,0) = \tfrac{1}{3}[4u(x,y,h/2) - u(x,y,h)] + O(h^4). \tag{4}$$

Thus we solve the problem twice and calculate (4) at the points common to each grid. Some typical results are given in Table II.

The result computed by series solution is

$$u(\tfrac{1}{4},\tfrac{1}{4}) - 0.4320283. \tag{5}$$

TABLE II

h	$u(x, y, h)$	$u(x, y, 0)$
$\frac{1}{8}$	0.43105	
$\frac{1}{16}$	0.43178	0.43202
$\frac{1}{32}$	0.43197	0.43202
$\frac{1}{64}$	0.43201 (27)	0.43202
$\frac{1}{128}$	0.43202 (44)	0.4320283

These results clearly indicate that the deferred approach to the limit technique yields better results than going to smaller step size, without significantly increasing the required time and storage. It bears repeating that this technique may not be easily applicable to the iterative techniques since, unless we do an excessive number of iterations, the errors may consist of more than local truncation.

13. General Linear Equations

Given a region R with boundary Γ the equation

$$u_{xx} + u_{yy} - g(x, y)\,u + \phi(x, y) = 0, \tag{1}$$

with the boundary conditions

$$u(x, y) = f(x, y) \quad \text{on} \quad \Gamma, \tag{2}$$

is the Euler equation associated with the minimization of the functional

$$J(u) = \int\!\!\int_R [u_x^2 + u_y^2 + g(x, y)\,u^2 - 2\phi(x, y)\,u]\,dx\,dy, \tag{3}$$

subject to (2).

If we perform the same discretization as in Section 2

$$\begin{aligned} \partial u_{ij}/\partial x &= [(u_{i+1,j} - u_{ij})/h] + O(h), \\ \partial u_{ij}/\partial y &= [(u_{i,j+1} - u_{ij})/h] + O(h), \end{aligned} \tag{4}$$

and define

$$g_{ij} = g(ih, jh)h^2, \qquad \varphi_{ij} = \varphi(ih, jh)h^2, \tag{5}$$

then the discretized version of (3) is

$$J(u) = \sum_{i=1}^{N}\Big[\sum_{j=1}^{M}(u_{ij} - u_{i,j-1})^2 + \sum_{j=1}^{m-1}[(u_{ij} - u_{i+1,j})^2 + g_{ij}u_{ij}^2 - 2\varphi_{ij}u_{ij}]\Big]. \tag{6}$$

Now if we define the matrix Q, vectors r_R and scalars s_r as before and further define the diagonal matrices G_R and vectors φ_R as

$$G_R = \operatorname{diag}(g_{R1}, g_{R2}, \ldots, g_{R,M-1}), \qquad \varphi_R = [\varphi_{Rj}], \tag{7}$$

then the inner product form of (6) is

$$\begin{aligned} J(u) = \sum_{R=1}^{N} [(Qu_R, u_R) - (2r_R, u_R) + s_R + (G_R u_R, u_R) - (2\varphi_R, u_R) \\ + (u_R - u_{R-1}, u_R - u_{R-1})]. \end{aligned} \tag{8}$$

We can now proceed in the same manner as before. We define

$$\begin{aligned} f_R(v) = \min_{u_R, u_{R+1}, \ldots, u_{N-1}} \sum_{i=R}^{N} [([Q + G_i]u_i, u_i) - (r_i + \varphi_i, 2u_i) \\ + (u_i - v, u_i - v) + s_i], \end{aligned} \tag{9}$$

with

$$v = u_{R-1}. \tag{10}$$

Using the principle of optimality, we obtain the recurrence relation

$$\begin{aligned} f_R(v) = \min_{u_R} [([Q + G_R]u_R, u_R) - (r_R + \varphi_R, 2u_R) + (u_R - v, u_R - v) \\ + s_R + f_{R+1}(u_R)], \qquad R = 1, 2, \ldots, N-1, \end{aligned} \tag{11}$$

subject to

$$f_N(v) = ([Q + G_N]u_N, u_N) - (r_N + \varphi_N, 2u_N) + (u_N - v, u_N - v) + s_N. \tag{12}$$

Again, we note that $f_R(v)$ can be written as

$$f_R(v) = (A_R v, v) - (2b_R, v) + c_R. \tag{13}$$

Thus we find that the minimizing u_R is

$$u_R = [I + Q + G_R + A_{R+1}]^{-1}(v + b_{R+1} + r_R + \varphi_R), \tag{14}$$

while the A_R and b_R satisfy the recurrence equations

$$\begin{aligned} A_R &= I - [I + Q + G_R + A_{R+1}]^{-1}, \\ b_R &= [I - A_R](r_R + \varphi_R + b_{R+1}), \end{aligned} \tag{15}$$

with the initial conditions

$$A_N = I, \qquad b_N = u_N. \tag{16}$$

We have again discarded the equation for c_R since we are only interested in the u_R. Thus, we solve (15) with the initial conditions, (16), and store the results. Then, we can rewrite (14) as

$$u_R = [I - A_R]\, u_{R-1} + b_R. \tag{17}$$

This is an initial value problem starting with u_0. The details are left to the reader.

If we go back through the proof of nonsingularity and stability, it should be clear that a *sufficient* condition to ensure both the nonsingularity of the matrices A_R and the stability of the method is

$$g(x, y) \geqslant 0. \tag{18}$$

This guarantees the positive definiteness of the quadratic functional.

14. Irregular Regions

We have already noted that if we have already solved a given equation over a particular region and retained the matrices A_R, we can solve the equation again with a new set of boundary conditions with very little effort. We can also note from Section 11 that the matrices A_R are independent of the forcing function $\varphi(x, y)$. The matrices A_R constitute the Green's function for the discrete problem and are functions of only the equation, without the forcing function, and the region.

Furthermore, it is clear that if we have the matrices $A_N, A_{N-1}, \ldots, A_1$, these are sufficient to solve problems over the truncated regions as shown in Fig. 2. For this region we will need $A_N, A_{N-1}, \ldots, A_{K+1}$. This argument indicates that in many applications it will be profitable to keep a file of the matrices A_R of various dimensions on tape.

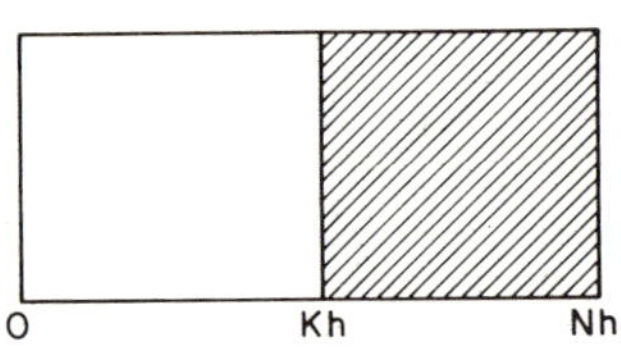

FIGURE 2

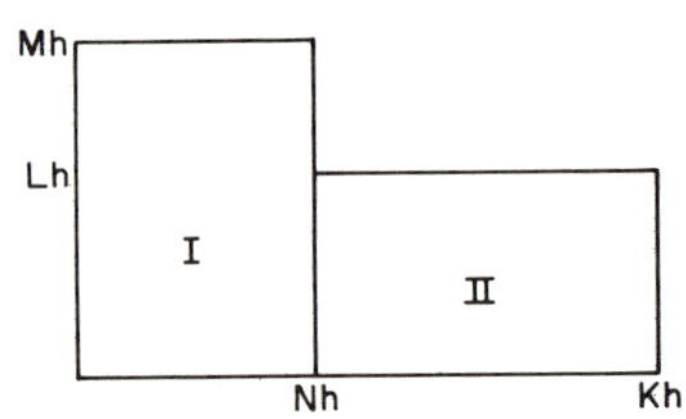

FIGURE 3

Now we would like to consider a building block technique for irregularly shaped regions. In Chapter 7 we will discuss a direct method for these regions. Suppose we have a region such as shown in Fig. 3. The region can be broken up into two rectangles, I and II. Let us assume that we are trying to solve the potential equation and we have already had occasion to solve the rectangular problems (I and II). If we further assume that the rectangular problem has been solved left to right in region I and right to left in region II, we then have already computed the solution of

$$A_R = I - [A_{R-1} + I + Q]^{-1}, \qquad R = 1, 2, \ldots, N-1, \quad A_0 = I, \tag{1}$$

where the matrices are of order $M-1$. We have also already solved

$$\bar{A}_R = I - [\bar{A}_{R+1} + I + Q]^{-1}, \qquad R = N+1, N+2, \ldots, K-1, \quad \bar{A}_K = I, \tag{2}$$

where these matrices are of order $L-1$. We would like to show that we can use these matrices to compute the solution over the combined region even though the boundary conditions may have changed.

For a given set of boundary conditions we can define r_R and $\bar{r}_R$ and then solve

$$b_R = [I - A_R](b_{R-1} + r_R), \qquad b_0 = u_0, \quad R = 1, 2, \ldots, N-1, \tag{3}$$

and

$$\bar{b}_R = [I - \bar{A}_R](\bar{b}_{R+1} + \bar{r}_R), \qquad \bar{b}_K = u_K, \quad R = N+1, \ldots, K-1, \tag{4}$$

since the orders of the quantities in each equation are consistent and u_0, u_K, r_R and $\bar{r}_R$ are given by the boundary conditions. The difficulty lies at u_N where the number of grid points changes. However, if we can determine u_N then we can obtain the solution in region I by means of the relation

$$u_R = [I - A_R]u_{R+1} + b_R, \qquad R = 1, 2, \ldots, N-1, \tag{5}$$

and the solution in region II using

$$u_R = [I - \bar{A}_R]u_{R-1} + \bar{b}_R, \qquad R = N+1, \ldots, K-1. \tag{6}$$

We will now show how u_N can be determined.

We can write u_N as

$$u_N = \begin{bmatrix} y \\ w \end{bmatrix}, \tag{7}$$

where

$$\begin{aligned} y &= (u_{N1}, u_{N2}, \ldots, u_{N,L-1}), \\ w &= (u_{N,L}, u_{N2,L+1}, \ldots, u_{N,M-1}). \end{aligned} \tag{8}$$

We want to make this distinction since y consist of interior points but w is given by the boundary conditions. Thus we need only determine y.

Using the definitions of Section 4, we denote the minimum value of the functional in region I by $f_{N-1}(u_N)$ and the minimum value of the functional in region II by $g_{N+1}(y)$. From the quadratic nature of the functional we have*

$$\begin{aligned} f_{N-1}(u_N) &= f_{N-1}\left(\begin{bmatrix} y \\ w \end{bmatrix}\right) = (A_{N-1} u_N, u_N) - (2b_{N-1}, u_N) + c_{N-1}, \\ g_{N+1}(y) &= (\bar{A}_{N+1} y, y) - (2\bar{b}_{N+1}, y) + \bar{c}_{N+1}. \end{aligned} \tag{9}$$

If $G(u)$ is the discrete functional over the entire region, then using additivity of the functional we see that

$$\min_{\{u_i\}} G(u) = \min_{y} \left\{ f_{N-1}\left(\begin{bmatrix} y \\ w \end{bmatrix}\right) + g_{N+1}(y) \right\}. \tag{10}$$

This is an application of the idea of the minimum convolution. Since $f_{N-1}\left(\begin{bmatrix} y \\ w \end{bmatrix}\right)$ and $g_{N+1}(y)$ are in a convenient form we can easily find y.

We partition A_{N-1} and b_{N-1} as follows,

$$A_{N-1} = \begin{bmatrix} A_{11} & A_{12} \\ A_{21} & A_{22} \end{bmatrix}, \qquad b_{N-1} = \begin{bmatrix} b_1 \\ b_2 \end{bmatrix}, \tag{11}$$

where A_{11} and b_1 have order $L-1$. Now we can expand (10) and obtain

$$\begin{aligned} \min_{\{u_i\}} G(u) = \min_{y} \big[&([A_{11} + \bar{A}_{N+1}] y, y) + ([A_{12} w + A_{21}^{\mathrm{T}}] w, y) - (2b_1, y) \\ &- (2\bar{b}_{N+1}, y)\big] - 2b_2, w) + (A_{22} w, w) + c_{N-1} + \bar{c}_{N+1}. \end{aligned} \tag{12}$$

This expression can be differentiated to find the minimizing y, which is, since A_{N-1} is symmetric,

$$y = [A_{11} + \bar{A}_{N+1}]^{-1} (A_{12} w - b_1 - \bar{b}_{N+1}). \tag{13}$$

---

* $\bar{c}_R$ and c_R should be defined as in Section 5. These quantities will not appear in the final expressions for u_R and need not be calculated.

Using this value of y, (5) and (6) are initial value problems starting with u_N and y respectively.

If then the matrices A_R and $\bar{A}_R$ are available, we first calculate b_R and $\bar{b}_R$, an easy task. Then we solve for y by (13) and finally solve (5) and (6) for the desired u_R. All of the operations require a small number of computations compared with the amount of effort required to solve the irregular region problem from scratch.

15. Higher Order Equations

Our technique is not restricted to second order linear equations. Consider, for example, the static deflection of an elastic plate under transverse loading as described by the biharmonic equation

$$u_{xxxx} + 2u_{xxyy} + u_{yyyy} = p, \tag{1}$$

a fourth order elliptic equation. The boundary conditions for (1) are usually given in terms of conditions at the edges of the plate. For a clamped plate we have

$$u(x,y) = 0, \tag{2}$$

at the edges, and a condition on the normal derivative

$$u_n(x,y) = 0, \tag{3}$$

at the edges.

It is well known, and easy to verify, that (1) is the Euler–Lagrange equation associated with the minimization of the functional

$$J(u) = \iint_R [(u_{xx} + u_{yy})^2 - 2pu]dx\, dy, \tag{4}$$

subject to (2) and (3). We can attempt to find approximate solutions of (1) by treating a discretized version of (4). Proceeding as before, we replace u_{xx}, u_{yy} and u_{xy} in (4) by finite difference approximations. In a direct manner we obtain a functional equation of the form

$$f_R(v, w) = \min_{u_R} [G(u_R, v, w) + f_{R+1}(u_R, w)], \tag{5}$$

where

$$v = u_{R-1}, \qquad w = u_{R-2}.$$

We can then simplify considerably by using the quadratic nature of (4) as before. We leave the details to the reader.

A simpler approach goes as follows. Let us define a function v by

$$v = u_{xx} + u_{yy}, \tag{6}$$

where u is a solution of (1). It is easy to verify that v must satisfy the equation

$$v_{xx} + v_{yy} = p. \tag{7}$$

Now if we define a vector variable w by

$$w = \begin{bmatrix} u \\ v \end{bmatrix},$$

we find from (6) and (7) that

$$w_{xx} + w_{yy} = Aw + s, \tag{8}$$

where

$$A = \begin{bmatrix} 0 & 1 \\ 0 & 0 \end{bmatrix}, \qquad s = \begin{bmatrix} 0 \\ p \end{bmatrix}.$$

Appropriate boundary condition for w can be found from (2) and (3). Since (8) is the Euler equation associated with the minimization of the functional

$$J(w) = \iint_R [(w_x, w_x) + (w_y, w_y) + (Aw, w) - 2(s, w)]\, dx\, dy, \tag{9}$$

we can easily solve a discrete version of (9) by the techniques developed in Section 11. The only difference will be that for the fourth order problem, the matrices and vectors will now be of order $2(M-1)$. We will carry out the details of this problem by invariant imbedding in the next chapter. It is easy to show that the method will always provide the solution and that it is computationally stable.

EXERCISES

1. Derive the matrix recurrence equation corresponding to (8).

2. Show that the necessary inverses always exist.

3. Derive the associated vector recurrence relation. What is the initial condition for this relation if we are solving the problem defined by (1), (2) and (3).

4. In many physical situations we are given bending moments and normal forces along a portion of the boundary. Consider for example a rectangular plate with the edge $x = 0$ subject to bending moments $f(y)$ and normal forces $g(y)$ while (2) and (3) hold on the remaining edges. Thus, we will have boundary conditions of the form

$$-u_{xx}(0,y) - \nu u_{yy}(0,y) = f(y),$$

$$u_{xxx}(0,y) + (2-\nu)\,u_{xyy}(0,y) = g(y),$$

on $x = 0$, where ν is the Poisson ratio, a physical constant of the material. Instead of (4) we must consider here the more general functional

$$J(u) = \iint_R [(u_{xx}+u_{yy})^2 - 2(1-\nu)(u_{xx}u_{yy}-u_{xy}^2) - 2pu]\,dx\,dy$$

$$+2\int gu\,dy + 2\int fu_x\,dy,$$

in order to consider forces f and g on an edge. Rederive the dynamic programming equations in terms of the variables

$$M(x,y) = -u_{xx} - \nu u_{yy}, \qquad V(x,y) = u_{xxx} + (2-\nu)\,u_{xyy}.$$

See the article by N. Distefano listed in the bibliography at the end of the chapter.

16. Distributed Control

Problems involving distributed control in general give rise to partial differential equations. Suppose we consider the problem of choosing a control function $v \equiv v(x,t)$ which minimizes

$$J(u,v) = \int_0^T \int_0^1 [g(x)u^2 + v^2]\,dx\,dt\,, \tag{1}$$

subject to

$$u_t = u_{xx} + v, \qquad u(x,0) = f(x), \qquad u(1,t) = u(0,t) = 0. \tag{2}$$

We suppose that $g(x) > 0$. We can easily verify that the Euler–Lagrange equations corresponding to (1) and (2) are

$$v_t = -v_{xx} + g(x)u, \qquad u_t = u_{xx} + v, \tag{3}$$

subject to

$$\begin{aligned} u(x,0) &= f(x), \qquad & u(1,t) &= u(0,t) = 0, \\ v(x,T) &= 0, & v(1,t) &= v(0,t) = 0. \end{aligned} \tag{4}$$

We will use dynamic programming plus discretization to treat problems (1) and (2).

The continuous problem can also be attacked directly, but it involves some more sophisticated concepts. Let us define

$$u_{ij} = u(ih, j\delta),$$

where

$$Mh = 1, \qquad N\delta = T.$$

We let u_i and v_i be respectively the $M-1$ dimensional vectors

$$u_i = [u_{ij}], \qquad j = 1, 2, \ldots, M-1,$$
$$v_i = [v_{ij}].$$

We can replace (1) and (2) by the discrete problem of finding

$$\min_{\{v_i\}} \sum_{i=R}^{N} [(Gu_i, u_i) + (v_i, v_i)], \tag{5}$$

subject to

$$u_{i+1} = [I - \rho Q] u_i + h v_i, \tag{6}$$

where Q is as defined previously,

$$G = \operatorname{diag}[g(h), \ldots, g(2h), g((M-1)h)],$$

and $\rho = h/\delta^2$. The initial condition for (6), u_0, is given by (4).

If we consider the sequence of variational problems

$$f_R(c) = \min_{\{v_i\}} \sum_{i=R}^{N} [(Gu_i, u_i) + (v_i, v_i)], \qquad u_R = c,$$

subject to

$$u_{i+1} = [I - \rho Q] u_i + h v_i,$$

we find proceeding as before that

$$\begin{aligned} f_R(c) &= \min_{v} [(Gc, c) + (v, v) + f_{R+1}([I - GQ] c + hv)], \\ f_N(c) &= (Gc, c). \end{aligned} \tag{7}$$

We see inductively that

$$f_R(c) = (A_R c, c).$$

We can now derive the minimizing v, namely

$$v = -h[I - h^2 A_{R+1}]^{-1} A_{R+1} [I - Q], \tag{8}$$

and a recurrence equation for A_R

$$A_R = G + [I - \rho Q]\,[I - [I + h^2 A_{R+1}]^{-1}]\,[I - \rho Q], \tag{9}$$

with

$$A_N = G.$$

We can show that under suitable restrictions on the ration ρ that the necessary matrix inverses always exist and that the method is stable. In Chapter 10 we will see that by changing our original discretization slightly we can ensure existence and stability for any positive value of ρ.

EXERCISES

1. Show that if we discretize the equation

$$u_t = u_{xx} + v$$

as

$$u_{i+1} = [I + \rho Q]^{-1} u_i + [I + \rho Q]^{-1} v_i,$$

the recurrence relation for A_R becomes

$$A_R = G + P[I - [I + h^2 P A_{R+1} P]^{-1}]P,$$

where

$$P = [I + \rho Q]^{-1}.$$

2. Show that this equation has a solution and is computationally stable for any positive value of ρ.

BIBLIOGRAPHY AND COMMENT

Section 2. This discretization is standard. It has long been noted that starting with the variational problem will lead to symmetric difference schemes. See

J. Todd, ed., *Survey of Numerical Analysis*, McGraw-Hill, New York, 1962.

Section 4. A number of discrete dynamic programming problems are considered in

R. Bellman, *Introduction to Matrix Analysis*, 2nd ed., McGraw-Hill, New York, 1970.

Section 5. The corresponding results for the semidiscrete, or method of lines, technique appear in

E. Angel, "Dynamic Programming and Linear Partial Differential Equations," *J. Math. Anal. Appl.* Vol. 23, 1968, pp. 628–638.

Section 7. In most references, the Perron–Frobenius theory of nonnegative matrices is employed to prove existence and stability. See, for example,

R. Varga, *Matrix Iterative Analysis*, Prentice-Hall, Englewood Cliffs, New Jersey, 1962.

Section 8. Ill-conditioning has not been a problem in practice with matrices of order less than 128.

Section 10. See

F. W. Dorr, "The Direct Solution of the Discrete Poisson Equation on a Rectangle," *SIAM Rev.*, Vol. 12, 1970, pp. 248–263.

Section 12. See

L. F. Richardson, "The Approximate Arithmetical Solution by Finite Differences of Physical Problems Involving Differential Equations, with Applications to the Stresses in a Masonry Dam," *Philos. Trans. Roy. Soc. London Ser. A.* Vol. 210, 1910, pp. 307–357.

Other convergence accelerating techniques can be found in

J. Todd, ed., *Survey of Numerical Analysis*, McGraw-Hill, New York, 1962.

Section 14. See

E. Angel, "A Building Block Technique for Elliptic Boundary-Value Problems Over Irregular Regions," *J. Math. Anal. Appl.*, Vol. 26, 1969, pp. 75–81.

This technique is similar to a tabular method suggested in

L. V. Kantorovich, V. I. Krylov, and K. Y. Chernin, Tables for the Numerical Solution of Boundary Value Problems of the Theory of Harmonic Functions, Ungar, New York, 1963.

Section 15. See

N. Distefano, "Dynamic Programming and the Solution of the Biharmonic Equation," *Internat. J. Numer. Meth. Engrg.*, Vol. 3, 1971, pp. 199–213.

Section 16. See

A. P. Sage, *Optimum Systems Control*, Prentice-Hall, Englewood Cliffs, New Jersey, 1968.

R. Bellman, *Introduction to the Mathematical Theory of Control Processes, II: Nonlinear Processes*, Academic Press, New York, 1971.

There are many interesting questions connected with the Dirichlet functional $D(u)$. See

R. Bellman and H. Osborn, "Dynamic Programming and the Variation of Green's Function," *J. Math Mech.*, Vol. 7, 1958, pp. 81–86, and other papers by Osborn.

Chapter 6

Invariant Imbedding

1. Invariant Imbedding

During the past ten years, the theory of invariant imbedding has been used to derive analytic and numerical results in a number of different fields, atmospheric physics, transport theory, wave propagation, to mention a few. From our point of view the invariant imbedding approach will allow us to replace linear two-point boundary value problems which often have associated unstable computational algorithms by initial value problems which possess simple stable algorithms. The method can be applied to cases where there are no associated variational processes.

We can give a motivation for the theory of invariant imbedding by considering the following problem.

Suppose we consider a neutron transport process in a rod of length L (Fig. 1). We will let $u(t)$ denote the flux of neutrons to the right at t, and $v(t)$ the flux to the left at t. The input fluxes, $u(0)$ and $v(L)$ are presumed given. If we consider the particle interactions between t and $t+\Delta$ where $\Delta \ll 1$, and then let $\Delta \to 0$, we can readily derive differential equations relating $u(t)$ and $v(t)$. In many cases, where there is no interaction between particles in the flux, these equations will take the simple form.

$$\begin{aligned} u' &= Au + Bv, \\ v' &= Cu + Dv, \\ u(0) = c_1, &\qquad v(L) = c_2. \end{aligned} \tag{1}$$

The two-point boundary value problem above is linear, and its solution can be made to depend on the solution of a linear system of algebraic equations. We suspect, however, that usual methods applied in (1) may be computationally unstable unless care is exercised.

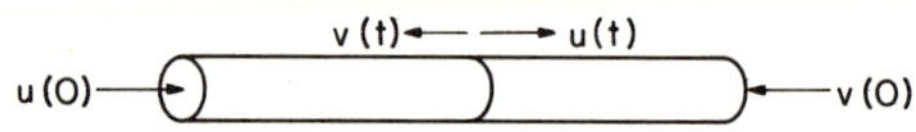

FIGURE 1

Suppose we now consider the problem from a different point of view. The flux to the right, $u(t)$, at some point of the rod consists of two components, the part of $u(0)$ which has been itself transmitted and the part of $v(t)$ which has been reflected back. If we let $T(t, L)$ and $R(t, L)$ represent the transmission and reflection coefficients respectively of a rod of length L at t, then

$$u(t) = T(t, L)\,u(0) + R(t, L)\,v(L). \tag{2}$$

A similar expression holds for $v(t)$. We see that (2) is a direct consequence of the linearity of (1).

If we consider how the reflection and transmission functions change as the length of the rod is increased from L to $L + \Delta L$, we can derive differential equations for $R(t, L)$ and $T(t, L)$ as functions of L. These equations will in general be stable initial value problems as far as most numerical methods are concerned, and thus can be easily solved numerically.

The method of invariant imbedding has been successfully applied to many classes of problems including ordinary differential, integral, and integro–differential equations. We shall be interested here mostly in difference equations.

2. The Riccati Transformation

Consider the coupled linear difference equations,

$$\begin{aligned} u_{i+1} &= A_i u_i + B_i v_i + e_i, \\ v_{i+1} &= C_i u_i + D_i v_i + f_i, \qquad i = 1, 2, \ldots, N-1, \end{aligned} \tag{1}$$

where u_i and v_i are M-dimensional vectors. We will assume that two-point boundary conditions

$$u_0 = c, \qquad v_N = d, \tag{2}$$

are specified.

We seek a solution to (1) in the form

$$v_i = R_i u_i + s_i, \tag{3}$$

where the matrix R_i and the vector s_i are independent of u_i and v_i. The existence of solutions of this form is a result of the linearity of (1). We will return to this point in the next chapter.

We can rewrite (3) as

$$v_{i+1} = R_{i+1} u_{i+1} + s_{i+1}, \tag{4}$$

and use (1) to relate R_{i+1} and s_{i+1} with R_i and s_i. Using (1) and (4) we have

$$C_i u_i + D_i v_i + f_i = R_{i+1}[A_i u_i + B_i v_i + e_i] + s_{i+1}, \tag{5}$$

and substituting for v_i with (3) we find

$$[C_i + D_i R_i] u_i + f_i + D_i s_i = R_{i+1}[A_i + B_i R_i] u_i + R_{i+1}(e_i + B_i s_i) + s_{i+1}. \tag{6}$$

Since this equation must hold for *all* u_i, we match coefficients in u_i to obtain the relations

$$\begin{aligned} [C_i + D_i R_i] &= R_{i+1}[A_i + B_i R_i], \\ f_i + D_i s_i &= R_{i+1}(e_i + B_i s_i) + s_{i+1}, \end{aligned} \tag{7}$$

or

$$\begin{aligned} R_i &= [R_{i+1} B_i - D_i]^{-1} [C_i - R_{i+1} A_i], \\ s_i &= [R_{i+1} B_i - D_i] [f_i - s_{i+1} - R_{i+1} e_i]. \end{aligned} \tag{8}$$

Since (3) holds for all u_i and v_i we have

$$d = R_N u_N + s_N. \tag{9}$$

This relation must hold regardless of the second boundary condition. Thus, we must have

$$R_N = 0, \qquad s_N = c. \tag{10}$$

We now have *initial value* problems for the R_i and s_i starting at $i = N$ and going backwards. To obtain an initial value problem for the u_i we use (3) in the first equation of (1) to get

$$u_{i+1} = [A_i + B_i R_i] u_i + e_i + B_i s_i . \tag{11}$$

This is an initial value problem for the determination of u_i starting with

$$u_0 = c \tag{12}$$

and using the values of R_i and s_i computed from (9) and (10). We will delay any discussion of existence of the necessary matrix inverses and of computational stability until we consider the particular values of A_i, B_i, C_i and D_i which will arise from the discretization of partial differential equations.

EXERCISES

1. Start with the representation

$$u_i = R_i v_i + s_i$$

and derive an initial value problem for the determination of the R_i and s_i.

2. Start with the representation $u(t) = R(t)v(t) + s(t)$ and derive initial value problems for the solution of (1.1).

3. Single Sweep Methods

We have established the fact that we can replace the coupled difference equations subject to two-point boundary conditions by two sets of initial value problems. Unfortunately, it appears that we have to store the results of the first problem, the determination of the R_i and s_i, to solve the second. It turns out, however, that under many circumstances the invariant imbedding approach will yield initial value problems without any storage requirement. We will illustrate this by means of a simple example.

Consider the *scalar* difference equations

$$\begin{aligned} u_{i+1} &= a_i u_i + b_i v_i , \\ v_{i+1} &= c_i u_i + d_i v_i , \end{aligned} \tag{1}$$

subject to the boundary conditions

$$u_0 = 0, \qquad v_N = 1 . \tag{2}$$

Since (1) is linear we can easily consider more general boundary conditions or add a forcing function using superposition. To emphasize that the process described by (1) and (2) has N stages, we rewrite (1) and (2) as

$$\begin{aligned} u_{i+1}(N) &= a_i u_i(N) + b_i v_i(N), \qquad u_0(N) = 0, \\ v_{i+1}(N) &= c_i u_i(N) + d_i v_i(N), \qquad v_N(N) = 1. \end{aligned} \tag{3}$$

Now suppose we consider the same equation subject to the same boundary conditions except that there are now $N+1$ stages,

$$\begin{aligned} u_{i+1}(N+1) &= a_i u_i(N+1) + b_i v_i(N+1), \qquad u_0(N+1) = 0, \\ v_{i+1}(N+1) &= c_i u_i(N+1) + d_i v_i(N+1), \qquad v_{M+1}(N+1) = 1. \end{aligned} \tag{4}$$

As a consequence of linearity, the solutions to (3) and (4) can differ only by some multiplicative constant k, i.e.,

$$\begin{aligned} u_i(N+1) &= k u_i(N), \\ v_i(N+1) &= k v_i(N). \end{aligned} \tag{5}$$

Setting $i = N$ in (5) and applying the boundary condition

$$v_N(N) = 1, \tag{6}$$

we see that

$$k = v_N(N+1) \tag{7}$$

or

$$\begin{aligned} u_i(N+1) &= v_N(N+1)\, u_i(N), \\ v_i(N+1) &= v_N(N+1)\, v_i(N). \end{aligned} \tag{8}$$

Thus we have our fundamental invariant imbedding relation, a description of how the solution changes as the process increases in duration.

Let us define

$$r_i = u_i(i). \tag{9}$$

If we let $i = N$ in (4) we find

$$\begin{aligned} r_{N+1} &= a_N u_N(N+1) + b_N v_N(N+1), \\ 1 &= c_N u_N(N+1) + d_N v_N(N+1). \end{aligned} \tag{10}$$

Setting $i = N$ in (8), we have

$$u_N(N+1) = v_N(N+1) r_N. \tag{11}$$

We can now solve (10) and (11) for r_{N+1} in terms of r_N

$$r_{N+1} = (a_N + b_N r_N)/(c_N + d_N r_N). \tag{12}$$

Since

$$u_0(N) = 0 \tag{13}$$

for all N, we have the initial condition for (12)

$$r_0 = 0. \tag{14}$$

If we return to the last section, we see that (12) is a scalar version of equation (7). However, we have still not used the full power of the invariant imbedding method.

Suppose we are interested in finding $u_k(N)$ and $v_k(N)$ for some specific values of k and N. We solve (10) for $v_N(N+1)$

$$v_N(N+1) = \frac{a_N - c_N r_{N+1}}{a_N d_N - b_N c_N}. \tag{15}$$

Next we use (15) in (8), obtaining

$$\begin{aligned} u_i(N+1) &= \frac{a_N - c_N r_{N+1}}{a_N d_N - b_N c_N} u_i(N), \\ & \qquad\qquad\qquad\qquad\qquad i \leqslant N. \\ v_i(N+1) &= \frac{a_N - c_N r_{N+1}}{a_N d_N - b_N c_N} v_i(N), \end{aligned} \tag{16}$$

To find $u_k(N)$ and $v_k(N)$ we proceed as follows. We solve (12) for $N = 1, 2, \ldots, k$. Then we adjoin (16) with the initial condition

$$u_k(k) = r_k, \qquad v_k(k) = 1, \tag{17}$$

and solve (12) and (16) for $N = k, k+1, \ldots, N-1$. Although there has been no storage required, we have obtained only the values $u_k(N)$ and $v_k(N)$. If we were also interested in the values $u_l(N)$ and $v_l(N)$, we would adjoin another set of equations like (16) to (12) and we would start the third set at $n = l$ with the initial conditions

$$u_l(l) = r_l, \qquad v_l(l) = 1. \tag{18}$$

Although this procedure is very useful, we will usually assume that we have sufficient storage to carry out our original procedure. However, the reader should keep in mind that the procedure we just described can be used as an alternative method and may in many instances prove superior.

EXERCISE

Derive the equivalent results for the vector equations of Section 2. Hint: Consider the matrix system

$$U_{i+1}(N) = A_i U_i(N) + B_i V_i(N),$$
$$V_{i+1}(N) = C_i U_i(N) + D_i V_i(N),$$
$$U_0(N) = 0, \qquad V_N(N) = I,$$

and show that

$$U_i(N+1) = U_i(N) V_N(N+1),$$
$$V_i(N+1) = V_i(N) V_N(N+1).$$

What physical interpretations do these equations have?

4. Discretization

We are now ready to return to the solution of elliptic partial differential equations. Once again as an example of the method we will use the potential equation

$$u_{xx} + u_{yy} = 0 \tag{1}$$

taken over the rectangle

$$0 \leqslant x \leqslant a, \qquad 0 \leqslant y \leqslant b, \tag{2}$$

with the boundary conditions given on all sides of the rectangle. We cover the region with a mesh and consider only the points

$$u_{ij} = u(ih, jh), \qquad i = 0, 1, \ldots, N, \qquad j = 0, 1, \ldots, M. \tag{3}$$

Again we have assumed that a and b are such that we can use a fixed grid spacing h. We replace the partial derivatives of (1) by the standard finite difference approximations

$$\begin{aligned} \partial^2 u_{ij}/\partial x^2 &= [(u_{i+1,j} - 2u_{ij} + u_{i-1,j})/h^2] + O(h^2), \\ \partial^2 u_{ij}/\partial y^2 &= [(u_{i,j+1} - 2u_{ij} + u_{i,j-1})/h^2] + O(h^2). \end{aligned} \tag{4}$$

We are thus led to solve the $(N-1)(M-1)$ linear difference equations

$$u_{i+1,j} + u_{i,j+1} - 4u_{ij} + u_{i,j-1} + u_{i-1,j} = 0, \qquad i = 1, 2, \ldots, N-1, \quad j = 1, 2, \ldots, M-1. \tag{5}$$

The values u_{i0}, u_{0j}, u_{Nj} and u_{iM} are given by the boundary conditions. Since the linear difference equations of (5) are "sparse" (no equation has more than five unknowns), iterative methods are usually used to solve the system. We will proceed in a different manner.

We define $M-1$ dimensional vectors u_R, an $M-1$ dimensional matrix Q, and $M-1$ dimensional vectors r_R exactly as we did in Chapter 5. Namely

$$u_R = [u_{Ri}],$$

$$Q = (q_{ij}) \qquad \text{where} \quad q_{ij} = \begin{cases} 2, & i = j, \\ -1, & |i-j| = 1, \\ 0, & \text{otherwise}, \end{cases} \tag{6}$$

$$r_R = [r_{Ri}] \qquad \text{where} \quad r_{Ri} = \begin{cases} u_{R0}, & i = 1, \\ u_{RM}, & i = M-1, \\ 0, & \text{otherwise}. \end{cases}$$

Now we can rewrite (5) as

$$u_{R+1} - 2u_R + u_{R-1} - Qu_R + r_R = 0, \tag{7}$$

where the boundary conditions are the two point conditions: u_0 and u_N known. One way to look at (7) is to observe that the first three terms approximate the x-derivatives for the vector u_R while the last two terms approximate the y-derivatives for this vector.

5. Recurrence Relations

Let us look for solutions of (4.7) of the form

$$u_{R+1} = A_R u_R + b_R, \tag{1}$$

where A_R and b_R are independent of u_R. The existence of such solutions was demonstrated in Section 2; see also the following chapter.

Substituting (1) into

$$u_{R+1} - 2u_R + u_{R-1} - Qu_R + r_R = 0, \tag{2}$$

and solving for u_R we find

$$u_R = [2I + Q - A_R]^{-1}(u_{R-1} + b_R + r_R). \tag{3}$$

Upon comparing (3) with (1) we have immediately

$$\begin{aligned} A_{R-1} &= [2I+Q-A_R]^{-1}, \\ b_{R-1} &= A_{R-1}(b_R+r_R). \end{aligned} \tag{4}$$

Considering (1) for $R = N-1$ we find the initial conditions for (4)

$$A_{N-1} = 0, \qquad b_{N-1} = u_N. \tag{5}$$

Again the calculation procedure is to solve the initial value problems (4), storing the results. Then the relation in (1) is an initial value problem starting with u_0.

6. Relation to Dynamic Programming

Since the computational procedure is similar for both the dynamic programming and Riccati transformation methods, it is interesting to look for a precise relation.

With A_R and b_R defined by (3.4) and (3.5), let us set

$$\begin{aligned} B_{R+1} &= I - A_R, \\ d_{R+1} &= b_R. \end{aligned} \tag{1}$$

Now if we use the recurrence equations in Section 3 we find equations for B_R and d_R. We have

$$\begin{aligned} B_R &= I - [I + Q - B_{R+1}]^{-1}, \\ d_R &= [I - B_R]\,(d_{R+1} + r_R), \end{aligned} \tag{2}$$

and

$$u_R = [I - B_R]\,u_{R-1} + d_R, \tag{3}$$

with the initial conditions

$$B_N = I, \qquad d_N = u_N. \tag{4}$$

These last three equations are exactly the relations we derived by dynamic programming.

7. Nonsingularity and Stability

Since the matrices derived by the Riccati transformation procedure, A_R, are related to those obtained using dynamic programming by the relation

$$A_R = I - B_{R+1}, \tag{1}$$

it is immediate that

$$0 < A_R < I, \qquad R < N-1 \tag{2}$$

since

$$0 < B_R < I, \qquad R < N, \tag{3}$$

as we know from Chapter 5, Section 7. Therefore it is clear that all the required matrices exist and that the necessary inverses can be taken.

To demonstrate stability we employ the same technique as before. Suppose that we are using the recurrence relation

$$A_{R-1} = [2I + Q - A_R]^{-1}, \tag{4}$$

and that an error is made. Then we may consider that we are actually solving

$$\bar{A}_{R-1} = [2I + Q - \bar{A}_R]^{-1}, \tag{5}$$

where

$$\bar{A}_R = A_R + E_R, \tag{6}$$

is the desired solution plus an error term. After carrying out some straightforward algebra we find as before that

$$E_{R-1} \cong [2I + Q - A_R]^{-1} E_R [2I + Q - A_R]^{-1}, \tag{7}$$

under the assumptions that the initial error is small. Thus, we have

$$E_{R-1} \cong A_{R-1} E_R A_{R-1}. \tag{8}$$

Taking norms in (8) and remembering that by (2)

$$\| A_{R-1} \| < 1, \tag{9}$$

we find

$$\| E_{R-1} \| < \| E_R \|, \tag{10}$$

and thus this procedure is computationally stable. Similar results hold for the equation

$$b_{R-1} = A_{R-1}(b_R + r_R), \tag{11}$$

and for the equation

$$u_{R+1} = A_R u_R + b_R. \tag{12}$$

We conclude that the method is computationally stable.

8. Relation to Gaussian Elimination

The standard finite difference equations of (4.5) can be written as the system of linear algebraic equations

$$Pw = d, \tag{1}$$

where P is a matrix of order $(N-1)(M-1)$ determined by the discretization, w is a vector of the unknown interior points and d is determined by the boundary conditions. Using the notation of Section 4 we can rewrite (1) in block tridiagonal form

$$\begin{bmatrix} [2I+Q] & -I & & & \bigcirc \\ -I & [2I+Q] & -I & & \\ & & \ddots & & \\ & & -I & [2I+Q] & -I \\ \bigcirc & & & -I & [2I+Q] \end{bmatrix} \begin{bmatrix} u_1 \\ u_2 \\ \vdots \\ u_{N-2} \\ u_{N-1} \end{bmatrix} = \begin{bmatrix} u_0 + r_1 \\ r_2 \\ \vdots \\ r_{N-1} \\ u_N + r_N \end{bmatrix}. \tag{2}$$

We can solve (2) by Gaussian elimination taking into account the blocks of zeros. We find that (5.4) corresponds to the triangularization of the system, while (5.1) corresponds to back substitution. The relation between triangularization, dynamic programming and invariant imbedding has been previously noted. The full power of the method we present becomes apparent in the solution of elliptic equations over irregular regions.

9. Relation to the Riccati Equation

We can solve (5.4) for A_R in terms of A_{R-1},

$$A_R = Q + 2I - A_{R-1}^{-1}, \tag{1}$$

or

$$A_R = Q + 2I - [I - (I - A_{R-1})]^{-1}. \tag{2}$$

Since the modulus of $I - A_{R-1}$ must be less than one by (7.2), we can expand $[I-(I-A_{R-1})]^{-1}$ in a geometric series

$$[I - (I - A_{R-1})]^{-1} = I + (I - A_{R-1}) + (I - A_{R-1})^2 + \cdots. \tag{3}$$

Then we can write the difference $A_R - A_{R-1}$ in the following form

$$A_R - A_{R-1} = Q - (I - A_{R-1})^2 - (I - A_{R-1})^3 \cdots . \tag{4}$$

Defining B_R as $I - A_R$, we have,

$$B_R - B_{R-1} \cong Q - B_{R-1}^2 . \tag{5}$$

This equation resembles the Riccati differential equation

$$dS/dx = Q - S^2 . \tag{6}$$

In fact the Riccati equation in (6) is precisely the one which arises from the continuous dynamic programming or invariant imbedding approach.

10. Invariant Imbedding

We can also solve our two-point boundary problem by invariant imbedding. We can either work with coupled difference equations as in Section 5 or we can proceed directly from the second order problem. We will take the later course.

Noting that the solution of our problem depends on the length of our region we write*

$$u(i+1, N) - [2I + Q]\, u(i, N) + u(i-1, N) + r(i) = 0 , \tag{1}$$

with

$$u(0, N) = c, \qquad u(N, N) = d. \tag{2}$$

As a consequence of linearity we can write the solution of (1) and (2) for any $i \leqslant N$ as

$$u(i, N) = U(i, N)\, d + p(i, N), \tag{3}$$

where the matrix $U(i, N)$ is the solution of

$$\begin{gathered} U(i+1, N) - [2I + Q]\, U(i, N) + U(i-1, N) = 0, \\ U(0, N) = 0, \qquad U(N, N) = I, \end{gathered} \tag{4}$$

and the vector $p(i, N)$ is the solution of

$$\begin{gathered} p(i+1, N) - [2I + Q]\, p(i, N) + p(i-1, N) + r(i) = 0, \\ p(0, N) = c, \qquad p(N, N) = 0. \end{gathered} \tag{5}$$

---

* In this section it will be convenient to use the notation $u(i, N)$ rather than $u_i(N)$.

We will solve this problem by seeing how the solutions change as a function of N.

First we consider (4) on interval lengths N and $N+1$ e.g., the problems

$$\begin{gathered} U(i+1,N) - [2I+Q]\,U(i,N) + U(i-1,N) = 0, \\ U(0,N) = 0, \qquad U(N,N) = I, \end{gathered} \tag{6}$$

and

$$U(i+1,N+1) - [2I+Q]\,U(i,N+1) + U(i-1,N+1) = 0\,, \tag{7}$$
$$U(0,N+1) = 0, \qquad U(N+1,N+1) = I.$$

It is clear by linearity that the solutions (6) and (7) can differ only by a multiplicative constant. Using the boundary conditions we find that

$$U(i,N+1) = U(i,N)\,U(N,N+1)\,. \tag{8}$$

This is our fundamental imbedding relationship. Let us define $R(N)$ by

$$R(N) = U(N,N+1)\,. \tag{9}$$

From the boundary condition we have

$$R(0) = 0\,. \tag{10}$$

With $i=N$, (7) becomes

$$I - [2I+Q]\,R(N) + U(N-1,N+1) = 0, \tag{11}$$

and with $i=N-1$, (8) becomes

$$U(N-1,N+1) = R(N-1)\,R(N)\,. \tag{12}$$

We can now eliminate $U(N-1,N+1)$ from (11) and (12), leaving a recurrence equation for $R(N)$

$$R(N) - [2I + Q - R(N-1)]^{-1}\,. \tag{13}$$

Thus to find $U(n,N)$ we solve

$$R(i) = [2I + Q - R(i-1)]^{-1}, \qquad R(0) = 0, \tag{14}$$

for $i = 1, 2, \ldots, n$. Then we adjoin (8) using (9)

$$U(n,i+1) = U(n,i)\,R(i)\,, \tag{15}$$

subject to

$$U(n,n) = R(n)\,. \tag{16}$$

Now we solve both (14) and (15) for $i = n, \ldots, N-1$.

We proceed in a similar manner to solve (5). Writing down the problem for interval lengths of N and $N+1$ we have

$$\begin{gathered} p(i+1,N) - [2I + Q]\,p(i,N) + p(i-1,N) + r(i) = 0, \\ p(0,N) = c, \qquad p(N,N) = 0, \end{gathered} \tag{17}$$

and

$$\begin{gathered} p(i+1,N+1) - [2I + Q]\,p(i,N+1) + p(i-1,N+1) + r(i) = 0, \\ p(0,N+1) = c, \quad p(N+1,N+1) = 0. \end{gathered} \tag{18}$$

Once again we use linearity to relate $p(i,N)$ and $p(i,N+1)$. If we look at the difference $p(i,N+1)-p(i,N)$ we see that this difference is zero for $i=0$. Therefore this difference must be a constant multiple of the solution to (4). We apply the boundary conditions to find this constant and arrive at the imbedding relationship

$$p(i,N+1) = p(i,N) + U(i,N)\,p(N,N+1). \tag{19}$$

We define $s(N)$ by

$$s(N) = p(N,N+1), \tag{20}$$

and thus from (17)

$$s(0) = c. \tag{21}$$

Setting $i=N$ in (18) we can write

$$-[2I + Q]\,s(N) + p(N-1,N+1) + r(N) = 0, \tag{22}$$

and with $i=N-1$ in (19) we have

$$p(N-1,N+1) = s(N-1) + R(N-1)\,s(N). \tag{23}$$

Combining (14), (22) and (23) we have the recurrence relation

$$s(N) = R(N)\,[s(N-1) + r(N)]. \tag{24}$$

We can now summarize the invariant imbedding method to find $u(n,N)$, the solution to (1) and (2) for some $n \leqslant N$. We start by solving

$$\begin{aligned} R(i) &= [2I + Q - R(i-1)]^{-1}, \\ s(i) &= R(i)\,[s(i-1) + r(i)], \end{aligned} \tag{25}$$

subject to

$$R(0) = 0, \qquad s(0) = c. \tag{26}$$

At $i = n$, we adjoin the equations

$$\begin{aligned} U(n, i+1) &= U n, i) R(i), \\ p(n, i+1) &= p(n, i) + U(n, i) s(i), \end{aligned} \tag{27}$$

subject to

$$\begin{aligned} U(n, n) &= I, \\ p(n, n) &= 0. \end{aligned} \tag{28}$$

We continue solving (25) and (27) until we have $U(n, N)$ and $p(n, N)$. Finally

$$u(n, N) = U(n, N) c + p(n, N). \tag{29}$$

It is clear for the invariant imbedding method that, while we are solving the same fundamental equations, we no longer have to store the $R(i)$ and $s(i)$. This is a major advantage. The disadvantage is that for each vector $u(k, N)$ we want, we must adjoin another set of equations similar to (27) starting at $i = k$.

EXERCISES

1. We have already shown that (25) has a unique solution and that the equations are computationally stable. Show that the same is true of (27).

2. Apply the invariant imbedding method to the system $u''(t) + g(t) u(t) - f(t)$, $u(0) = c_1$, $u(\tau) = c_2$.

11. Continuous Invariant Imbedding

We could have approached the subject of obtaining numerical solutions of elliptic boundary value problems from another point of view. We could have kept all the variables continuous and looked for continuous initial value problems. As an example, we will consider the following problem.

Let $u(x, y)$ be the solution of

$$u_{xx} + u_{yy} = 0 \tag{1}$$

on the rectangle

$$0 \leqslant x \leqslant a, \qquad 0 \leqslant y \leqslant 1,$$

subject to the boundary conditions

$$u(0, y) = u(x, 0) = u(x, 1) = 0, \tag{2}$$

and the new boundary condition

$$u_x(a, y) = g(y). \tag{3}$$

To emphasize the dependence of u on the length of the rectangle we will write

$$u(x, y) = u(x, y, a). \tag{4}$$

Let $v(x, y, a, s)$ be the solution of

$$v_{xx} + v_{yy} = 0, \tag{5}$$

on the same rectangle with the boundary conditions

$$v(0, y, a, s) = v(x, 0, a, s) = v(x, 1, a, s) = 0 \tag{6}$$

and

$$v_x(a, y, a, s) = \delta(y - s), \qquad 0 < s < 1, \tag{7}$$

where $\delta(y)$ is the Dirac delta function. Thus, the function $v(x, y, a, s)$ is intimately related to the Green's function. The use of the delta function can readily be legitimized by means of the theory of distributions. Let us proceed formally here.

Comparing the systems for u and v, we have by superposition

$$u(x, y, a) = \int_0^1 v(x, y, a, z) g(z)\, dz. \tag{8}$$

Hence if we can determine v, the desired problem will be solved.

To find v, we start by differentiating (5)–(7) with respect to the interval length, a. We find that the function v_3 satisfies*

$$(v_3)_{11} + (v_3)_{22} = (v_3)_{xx} + (v_3)_{yy} = 0, \tag{9}$$

and from (6) we obtain the boundary conditions

$$v_3(0, y, a, s) = v_3(x, 0, a, s) = v_3(x, 1, a, s) = 0. \tag{10}$$

We get the final boundary condition by differentiation of (7) with respect to a yielding the equation

$$v_{11}(a, y, a, s) + v_{13}(a, y, a, s) = 0. \tag{11}$$

---

* To avoid confusion we will use the notation $v_i, i = 1, 2, 3, 4$ to denote differentiation of v with respect to the ith variable. Thus, $v_3(x, y, a, s) = v_a(x, y, a, s)$.

Comparing (9)–(11) with (5)–(7) we can apply superposition to obtain our fundamental imbedding equation

$$v_3(x, y, a, s) = -\int_0^1 v(x, y, a, z) v_{11}(a, z, a, s)\, dz\,. \tag{12}$$

Since v is a solution of the potential equation, we can rewrite (12) as

$$v_3(x, y, a, s) = \int_0^1 v(x, y, a, z) v_{22}(a, z, a, s)\, dz\,. \tag{13}$$

Let us define $r(a, w, s)$ by

$$r(a, w, s) = v(a, w, a, s)\,. \tag{14}$$

From (6) we have the initial and boundary conditions

$$r(0, w, s) = 0\,, \qquad r(a, 0, s) = r(a, 1, s) = 0\,. \tag{15}$$

To find a differential equation for r we differentiate with respect to a

$$r_a(a, w, s) = v_1(a, w, a, s) + v_3(a, w, a, s)\,, \tag{16}$$

which becomes using (7)

$$r_a(a, w, s) = \delta(w - s) + v_3(a, w, a, s)\,. \tag{17}$$

Finally we replace v_3 by (13) evaluated at $x = a$ to obtain

$$r_a(a, w, s) = \delta(w - s) + \int_0^1 r(a, w, z) r_{zz}(a, z, s)\, dz\,. \tag{18}$$

To find $u(x, y, s)$, we solve (18) for $0 \leqslant a \leqslant x$ starting with (15). At $a = x$ we adjoin (13), which in view of (14) can be written as

$$v_3(x, y, a, s) = \int_0^1 v(x, y, a, z) r_{22}(a, z, s)\, dz\,, \tag{19}$$

which is an initial value problem starting with

$$v(x, y, x, s) = r(x, y, s)\,. \tag{20}$$

Finally at the desired l, we perform the integration in (8) to obtain $u(x, y, l)$.

A number of the standard numerical methods can be derived from (18). For instance if we discretize the variables w and s, that is,

$$w_i = ih, \qquad s_j = jh\,, \tag{21}$$

we will have a method of lines technique. If we further use the trapezoidal rule to approximate to the integrals in (18) we will have the standard method of lines technique. If we now discretize a, we will be left with a finite difference approach. The advantage of starting with (18) is that this equation suggests many new approaches. For example, we can use a Gaussian quadrature formula in (18). We would expect that the resulting method would be more accurate than the standard methods. Much, however, remains to be done in this area.

EXERCISES

1. Extend the above result to the potential equation with the boundary conditions

$$u(0,y) = f(y), \qquad u_x(a,y) = g(y), \qquad u(x,0) = a(x), \qquad u(x,1) = b(x).$$

2. Show that the discretization suggested in (21) and the trapezoidal rule, convert (18) to the matrix Riccati equation

$$R_a = I - (1/h^2)\, RQR.$$

3. Solve the problem defined by (1), (2) and

$$u(a,y) = f(y).$$

Hint: Use the solution for v and then solve a Fredholm equation at $x = a$.

4. Prove

$$r(x,y,z) = r(x,z,y).$$

12. Generalized Riccati Transformations

We can also generalize the Riccati transformation to the original continuous problem. Again let us consider the potential equation

$$u_{xx} + u_{yy} = 0, \tag{1}$$

on the rectangle $0 \leqslant x \leqslant a, 0 \leqslant y \leqslant b$, where the values of u are specified on all sides of the rectangle. We will seek solutions of the form

$$u(x,y) = \int_0^b r(x,y,z)\, u_x(x,z)\, dz + s(x,y). \tag{2}$$

Differentiation of (2) with respect to x yields

$$u_x(x,y) = \int_0^b r_x(x,y,z)\,u_x(x,z)\,dz + \int_0^b r(x,y,z)\,u_{xx}(x,z)\,dz + s_x(x,z)\,, \quad (3)$$

and, in view of (1),

$$\begin{aligned} u_x(x,y) = & \int_0^b r_x(x,y,z)\,u_x(x,z)\,dz \\ & - \int_0^b r(x,y,z)\,u_{zz}(x,z)\,dz + s_x(x,z)\,. \end{aligned} \quad (4)$$

Differentiating (2) with respect to y, we have

$$u_{yy}(x,y) = \int_0^b r_{yy}(x,y,z)\,u_x(x,z)\,dz + s_{yy}(x,y)\,. \quad (5)$$

Finally use of (5) and the sifting property of the delta function allows us to rewrite (4) as

$$\begin{aligned} \int_0^b \delta(y-z)\,u_x(x,z)\,dz = & \int_0^b r_x(x,y,z)\,u_x(x,z)\,dz \\ & - \int_0^b \int_0^b r(x,y,w)\,r_{ww}(x,w,z)\,dw\,dz \\ & - \int_0^b r(x,y,z)\,s_{zz}(x,z)\,dz + s_x(x,y)\,. \end{aligned} \quad (6)$$

Equating coefficients of u_x we have

$$\begin{aligned} r_x(x,y,z) &= \delta(y-z) + \int_0^b r(x,y,w)\,r_{ww}(x,w,z)\,dw\,, \\ s_x(x,y) &= \int_0^b r(x,y,z)\,s_{zz}(x,z)\,dz\,. \end{aligned} \quad (7)$$

If we require (2) to hold at the boundaries, we must have

$$r(x,y,z) = 0\,, \qquad s(x,y) = u(x,y)\,, \quad (8)$$

at $x = a$, $y = 0$ and $y = b$, which gives initial and auxiliary equations to determine r and s. Thus, r and s are determined backwards, i.e., from $x = a$ to $x = 0$. Since $u(0,y)$ is known, (2) gives us a Fredholm equation for the missing initial condition, $u_x(0,y)$, using $r(0,y,z)$ and $s(0,y)$. Finally combining (1) and (5), we find an initial value problem for u_x

$$u_{xx}(x, y) = -\int_0^b r_{yy}(x, y, z)\, u_x(x, z)\, dz + s_{yy}(x, y), \tag{9}$$

which uses the stored values of r and s. The desired values of u are given by (2).

13. The Biharmonic Equation

The biharmonic equation can also be treated by the continuous method with surprising results. For example, if we consider the deflections of a plate clamped on three sides, we wish to solve

$$w_{xxxx} + 2w_{xxyy} + w_{yyyy} = 0, \tag{1}$$

subject to

$$w = w_n = 0, \tag{2}$$

on the clamped edges, and

$$w_{xx} = f(y), \qquad w_{xxx} + 2w_{xyy} = g(y), \tag{3}$$

on the free edge, $x = a$. To solve (1)–(3) we introduce the functions $u(x, y, z)$ and $v(x, y, z)$ which satisfy (1) and (2), i.e.,

$$\begin{aligned} u_{xxxx} + 2u_{xxyy} + u_{yyyy} &= 0, \\ v_{xxxx} + 2v_{xxyy} + v_{yyyy} &= 0, \end{aligned} \tag{4}$$

with

$$u = u_n = v = v_n = 0, \tag{5}$$

on the edges $x = 0$, $y = 0$ and $y = 1$. Along the edge $x = a$, we require

$$\begin{aligned} u_{xx} &= \delta(y-z), \\ u_{xxx} + 2u_{xyy} &= 0, \\ v_{xx} &= 0, \\ v_{xxx} + 2v_{xyy} &= -\delta(y-z). \end{aligned} \tag{6}$$

Thus, w is given by superposition

$$w(x, y) = \int_0^1 u(x, y, z) f(z)\, dz - \int_0^1 v(x, y, z) g(z)\, dz. \tag{7}$$

We can now consider u and v also to be functions of the length of the plate a. We will leave the details to the reader. The importance of this approach is as follows. If we proceed as in Section 11 we find that the functions p, q, r, and s, which are defined as the deflections at the edge $x = a$,

$$\begin{aligned} p(x,y,z) &= u_1(x,y,x,z), \\ q(x,y,z) &= v_1(x,y,x,z), \\ r(x,y,z) &= u(x,y,x,z), \\ s(x,y,z) &= v(x,y,x,z), \end{aligned} \tag{8}$$

satisfy the integro–differential equations

$$\begin{aligned} p_a(a,x,y) &= \delta(x-y) + 2\int_0^1 p(a,y,z)p_{zz}(a,z,y)\,dz \\ &\quad - \int_0^1 q(a,x,z)r_{zzzz}(a,z,y)\,dz\,, \\ q_a(a,x,y) &= p(a,x,y) + 2\int_0^1 p(a,x,z)q_{zz}(a,z,y)\,dz \\ &\quad - \int_0^1 q(a,x,z)s_{zzzz}(a,z,y)\,dz\,, \\ r_a(a,x,y) &= p(a,x,y) + 2\int_0^1 r(a,x,z)p_{zz}(a,z,y)\,dz \\ &\quad - \int_0^1 s(a,x,z)r_{zzzz}(a,z,y)\,dz\,, \\ s_a(a,x,y) &= q(a,x,y) + r(a,x,y) + 2\int_0^1 r(a,x,z)q_{zz}(a,z,y)\,dz \\ &\quad - \int_0^1 s(a,x,z)s_{zzzz}(a,z,y)\,dz\,. \end{aligned} \tag{9}$$

It is not difficult to show that these equations together with the proper initial and auxiliary conditions have the property that

$$p(a,x,y) = p(a,y,x), \qquad s(a,x,y) = s(a,y,x), \qquad q(a,x,y) = r(a,y,x). \tag{10}$$

These symmetry properties are known as the Betti reciprocity relations and have great importance in structural mechanics.

EXERCISES

1. Start with u and v as defined by (4)–(6). Let

$$u = u(x, y, a, z), \qquad v = v(x, y, a, z).$$

Find initial value problems for u and v as functions of a.

2. Using the results of Exercise 1, prove the reciprocity relations of (10).

14. Random Walk

Let us now briefly discuss a fundamental relationship between potential theory and probability theory which will provide a simple interpretation of some of the results we have obtained by means of invariant imbedding.

FIGURE 2

Consider, to begin, a one-dimensional random walk problem (Fig. 2). A particle at i is assumed to have equal probability of going to the right or to the left, where i assumes the values $N+1, \ldots, M-1$. The points N and M are absorbing barriers in the sense that the process ends as soon as the particle reaches either point.

Introduce the function

$$v(i) = \text{the probability of reaching } M \text{ before } N \text{ upon starting at } i, \quad N+1 \leqslant i \leqslant M-1. \tag{1}$$

We can extend the function with the aid of the boundary conditions

$$v(N) = 0, \qquad v(M) = 1. \tag{2}$$

It follows from the foregoing that $v(i)$ satisfies the difference equation

$$v(i) = \tfrac{1}{2}[v(i+1) + v(i-1)], \tag{3}$$

or

$$v(i+1) - 2v(i) + v(i-1) = 0. \tag{4}$$

We recognize (5) as a discrete version of the one-dimensional potential equation

$$d^2v/dx^2 = 0, \qquad v(a) = 0, \quad v(b) = 1. \tag{5}$$

Let us now consider a two-dimensional random walk process defined on the grid points (m,n), $M_1 \leqslant m \leqslant M_2$, $N_1 \leqslant n \leqslant N_2$, where a particle has equal probability of going in any of the four directions (Fig. 3).

FIGURE 3

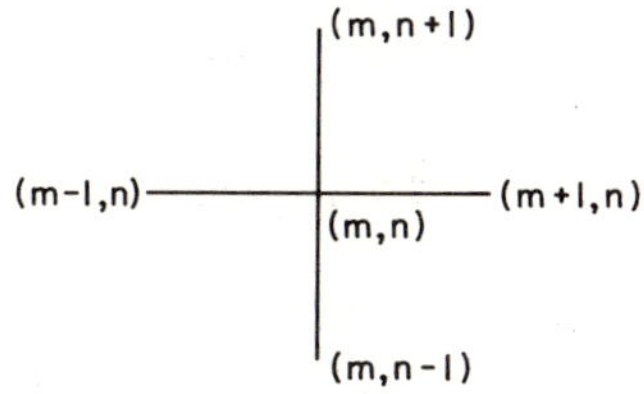

Introduce the function

$$u(m,n) = \text{the probability of reaching the edge } (M_2, n) \text{ before reaching any of the other edges, starting at } (m,n). \quad (6)$$

We see that $u(m,n)$ satisfies the relation

$$u(m,n) = \tfrac{1}{4}[u(m-1,n) + u(m+1,n) + u(m,n+1) + u(m,n-1)], \quad (7)$$

which is precisely the discretized version of the two-dimensional potential equation,

$$u_{xx} + u_{yy} = 0, \quad (8)$$

that we have been using.

15. Invariant Imbedding and Random Walk

Let $M+1 \geqslant j \geqslant i$, and introduce the function

$$u(i,j) = \text{the probability of reaching } j \text{, starting at } i \text{, before reaching } N. \quad (1)$$

FIGURE 4

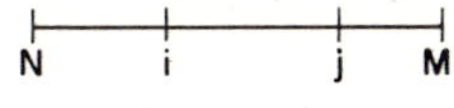

It is clear from Fig. 4 that

$$u(i,j+1) = u(i,j)\,u(j,j+1). \quad (2)$$

This result is a scalar version of (10.8). Similarly, we derive the matrix version, upon considering Fig. 5.

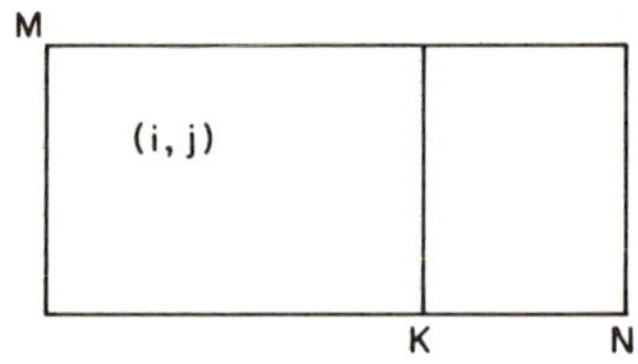

FIGURE 5

16. Another Imbedding

In this section we give another example of a nonclassical imbedding procedure. Although the technique is simple, it will illustrate one of the many ways the notions of imbedding can be used to derive new computational methods.

Suppose u satisfies the equation

$$u_{xx} + u_{yy} + \lambda g(x,y)\,u = 0\,, \tag{1}$$

in a region R, where

$$u = f(x,y)\,, \tag{2}$$

on the boundary Γ. In many cases λ is a parameter of physical importance. We will assume

$$\lambda \geqslant 0\,, \tag{3}$$

and we would like to investigate how the solution changes as a function of λ, i.e.,

$$u = u(x,y,\lambda)\,. \tag{4}$$

Differentiating (1) and (2) with respect to λ we have

$$(u_\lambda)_{xx} + (u_\lambda)_{yy} + \lambda g(x,y)\,u_\lambda = -\,g(x,y)\,u\,, \tag{5}$$

subject to

$$u_\lambda = 0\,, \tag{6}$$

on Γ. Let $G(x,y,a,b,\lambda)$ be the Green's function for (5) and (6). Hence u_λ is given by

$$u_\lambda(x,y,\lambda) = -\iint_R g(a,b)\,u(a,b)\,G(a,b,x,y,\lambda)\,da\,db\,. \tag{7}$$

Proceeding formally we can define the Green's function as the solution of

$$G_{xx}(x,y,a,b,\lambda) + G_{yy}(x,y,a,b,\lambda) + \lambda g(x,y)\,G(x,y,a,b,\lambda) = \delta(x-a,y-b)\,, \tag{8}$$

subject to

$$G(x, y, a, b, \lambda) = 0\,, \tag{9}$$

on Γ. Differentiating (8) and (9) with respect to λ we find

$$(G_\lambda)_{xx} + (G_\lambda)_{yy} + \lambda g(x,y)\,G_\lambda = -\,g(x,y)\,G\,, \tag{10}$$

subject to

$$G_\lambda = u\,, \tag{11}$$

on Γ. Thus G_λ is given by

$$G_\lambda(x, y, a, b, \lambda) = -\int_R\!\!\int g(a_1, b_1)\,G(a_1, b_1, a, b, \lambda)\,G(x, y, a_1, b_1, \lambda)\,da_1\,db_1\,. \tag{12}$$

We can now obtain $u(x, y, \lambda)$ as a function of λ as follows. For $\lambda = 0$ we solve

$$u_{xx} + u_{yy} = 0\,, \tag{13}$$

subject to (2), and

$$G_{xx} + G_{yy} = \delta(x-a, y-b)\,, \tag{14}$$

subject to (9). We now can solve (7) and (12) as initial value problems in λ subject to (2) and (9).

EXERCISE

Consider the problem of solving

$$u_{xx} + u_{yy} = g(x, y)\,,$$

subject to $u = \lambda f$ on Γ. Consider u as a function of λ.

BIBLIOGRAPHY AND COMMENT

Section 1. Invariant Imbedding was originally derived as an extension of the principles of invariance due to Ambarzumian and Chandrasekhar. See

V. A. Ambarzumian, "Diffuse Reflection of Light by a Foggy Medium," *Dokl. Akad. Nauk SSSR*, Vol. 38, 1943, pp. 229–232.

S. Chandrasekhar, *Radiative Transfer*, Dover, New York, 1960.

For further references and applications see

G. M. Wing, *Introduction to Transport Theory*, Wiley, New York, 1962.

R. Bellman, *Vistas of Modern Mathematics*, Univ. of Kentucky Press, Lexington, Kentucky, 1968.

R. Bellman and G. M. Wing, *Invariant Imbedding*, (to appear).

Section 2. The Riccati transformation has also been called the sweep method. Unfortunately, the latter notation has led to a great deal of confusion between the two-sweep method of the Riccati transformation and the single sweep method of invariant imbedding which we develop in Section 3. See

S. K. Godunov and V. S. Ryabenki, *Theory of Difference Schemes*, North-Holland Publ., Amsterdam, 1964.

I. M. Gelfand and S. V. Fomin, *Calculus of Variations*, Prentice-Hall, Englewood Cliffs, New Jersey, 1963.

G. B. Rybicki and P. D. Usher, "The Generalized Riccati Transformation as a Simple Alternative to Invariant Imbedding," *Astrophys. J.*, Vol. 146, 1966, pp. 871–879.

G. H. Meyer, "The Invariant Imbedding Equations for Multipoint Boundary-Value Problems," *SIAM J. Appl. Math.*, Vol. 18, 1970, pp. 433–453.

Section 3. See

R. Kalaba, *A One Sweep Method for Linear Difference Equations with Two Point Boundary Conditions*, Electronic Sciences Laboratory, Univ. of Southern California, USCEE 23, 1969.

E. Angel and R. Kalaba, "A One Sweep Numerical Method for Vector–Matric Difference Equations with Two-Point Boundary Conditions," *J. Optim. Theory Appl.*, Vol. 6, 1970, pp. 345–355.

R. Bellman, H. Kagiwada, and R. Kalaba, "Invariant Imbedding and the Numerical Integration of Boundary Value Problems for Unstable Linear Systems of Ordinary Differential Equations," *Comm. ACM*, Vol. 10, 1967, pp. 100–102.

Section 4. This discretization is the standard finite difference approximation. See any of the references for Chapter 4, Section 9, for convergence proofs.

Section 5. Solutions of the form given here have been long noted. See

A. F. Cornock, "The Numerical Solution of Poisson's and the Bi-Harmonic Equations by Matrices," *Proc. Camb. Phil. Soc.*, Part 4, Vol. 50, 1954, pp. 524–535.

O. Karlquist, "Numerical Solution of Elliptic Difference Equations by Matrix Methods," *Tellus*, Vol. 4, 1952, pp. 374–384.

An excellent bibliography is contained in the review article

F. W. Dorr, "The Direct Solution of the Discrete Poisson Equation on a Rectangle," *SIAM Review*, Vol. 12, 1970, pp. 248–263.

Section 7. The stability problem for other direct methods is discussed in

J. B. Rosser, "The Direct Solution of Difference Analogs of Poisson's Equation," Univ. of Wisconsin, Mathematics Research Center Tech. Summary Rept. 797, 1967.

Section 8. See

D. U. Von Rosenberg, *Methods for the Numerical Solution of Partial Differential Equations*, Amer. Elsevier, New York, 1969.

R. S. Lehman, "Dynamic Programming and Gaussian Elimination," *J. Math. Anal. Appl.*, Vol. 15, 1962, pp. 499–501.

W. R. Soillers and N. Hickerson, "Optimal Elimination for Sparse Symmetric Systems as a Graph Problem," *Quart. Appl. Math.*, Vol. 26, 1968, pp. 425–432.

Section 9. See

E. Angel, "Discrete Invariant Imbedding and Elliptic Boundary-Value Problems over Irregular Regions," *J. Math. Anal. Appl.* Vol. 23, 1968, pp. 471–484.

Section 10. See

E. Angel, "Invariant Imbedding and Partial Differential Equations," Proc. Summer Workshop on Invariant Imbedding 1970, R. Bellman and E. Denman, eds., Springer-Verlag, Berlin and New York, 1971.

Section 11. See

E. Angel, A. Jain and R. Kalaba, *Initial Value Methods in Potential Theory*, Electronic Sciences Laboratory, Univ. of Southern California, USCEE 22, 1970.

P. B. Bailey, "A Rigorous Derivation of Some Invariant Imbedding Equations of Transport Theory," *J. Math. Anal. Appl.* Vol. 8, 1964, pp. 144–169.

See also the paper by G. H. Meyer listed above and

M. Goldberg, "A Generalized Invariant Imbedding Equation," *J. Math. Anal. Appl.*, Vol. 34, 1971, pp. 590–601.

Section 12. See

E. Angel and A. Jain, "Initial-Value Transformation for Elliptic Boundary Value Problems," *J. Math. Anal. Appl.*, (to appear).

Other transformations have been suggested. See

C. Maynard and M. Scott, "Invariant Imbedding and Linear Partial Differential Equations via Generalized Riccati Transformations," *J. Math. Anal. Appl.*, (to appear).

Section 13. See

E. Angel, N. Distefano, and A. Jain, "Invariant Imbedding and the Reduction of Boundary-Value Problems of Thin Plate Theory to Cauchy Formulations," *Int. J. Eng. Sci.*, (to appear).

Section 14. See

R. Bellman and R. Kalaba, "Random Walk, Scattering and Invariant Imbedding I: One Dimensional Discrete Case," *Proc. Nat. Acad. Sci. U.S.*, Vol. 43, 1957, pp. 930–933.

Section 16. For ordinary differential equations, see

R. Huss and R. Kalaba, "Invariant Imbedding and the Numerical Determination of Green's Functions," *J. Opt. Theory Appl.*, (to appear).

In the study of Fredholm equations, these ideas have been used extensively; for instance the Bellman–Krein equation for the Fredholm resolvent uses the interval length as a parameter. See

R. Bellman, "Functional Equations in the Theory of Dynamic Programming—VIII, A Partial Differential Equation for the Fredholm Resolvent," *Proc. Amer. Math. Soc.*, Vol. 8, 1957, pp. 435–440.

The abstract structure behind these results can be found in

S. Bergman and M. Schiffer, *Kernal Functions and Elliptic Differential Equations in Mathematical Physics*, Academic Press, New York, 1953.

A. McNabb and A. Schumitzky, "Factorization of Operators I; Algebraic Theory and Examples," (to appear).

A. McNabb and A. Schumitzky, "Factorization and Operators III, Initial Value Methods for Linear Two-Point Boundary Value Problems," *J. Math. Anal. Appl.*, Vol. 31, 1970, pp. 391–406.

An application to a problem in structural mechanics is given in

E. Angel and N. Distefano, "Invariant Imbedding and the Effect of Changes in Poisson's Ratio in Thin Plate Theory," *Int. J. Eng. Sci.*, (to appear).

Chapter 7

Irregular Regions

1. Introduction

In the previous chapters we considered rectangular regions where it can be expected that many methods will work well. In this chapter we wish to consider some irregular regions which constitute a serious challenge to most methods.

2. Irregular Regions

We will consider a general region with the following two restrictions. First, all the grid points used in the discretization will be assumed to be regular points, that is, the spacing between adjacent points, including boundary points, will be the same. Second, the region will be assumed to have no holes in it in the following sense. Since our method involves going across the region in a particular direction (the x-direction), we require that the region can be oriented so that any line drawn through the region in the other direction (the y-direction) lies entirely in the region. Thus, a typical admissible region will have the form shown in Fig. 1.

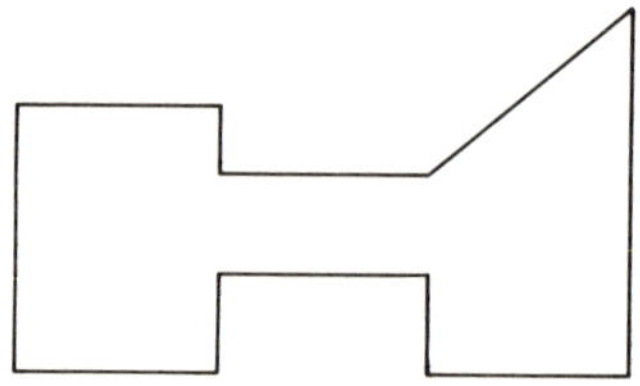

FIGURE 1

Neither of the two restrictions is necessary. However, we impose them since we would like to keep the discussion as simple as possible. We will discuss the removal of the restrictions later.

The method we will use to solve the potential equation over an irregular region is to see how (6.3.1) and (6.3.4) must be modified when the orders of two adjacent vectors of unknowns, u_R and u_{R-1}, are not the same. If u_R and u_{R-1} have the same order, equations (6.3.1) and (6.3.4) apply directly where the order of all matrices and vectors is that of u_R and u_{R-1}.

3. Case I: Order $u_R >$ Order u_{R-1}

It is sufficient to consider the situation shown in Fig. 2 where

$$\begin{aligned} u_R &= (u_{R1}, u_{R2}, \ldots, u_{RM}), \\ u_{R-1} &= (u_{R-1,1}, u_{R-1,2}, \ldots, u_{R-1,N}), \qquad M > N, \end{aligned} \tag{1}$$

and

$$w = (w_{N+1}, w_{N+2}, \ldots, w_M), \tag{2}$$

is a vector of known quantities determined by the boundary conditions. The case in which the lower boundaries of u_R and u_{R-1} do not match can be considered in exactly the same manner as the top boundaries.

Let $\tilde{u}_{R-1}$ be the vector

$$\tilde{u}_{R-1} = \begin{bmatrix} u_{R-1} \\ w \end{bmatrix} = (u_{R-1,1}, \ldots, u_{R-1,N}, w_{N+1}, \ldots, w_M). \tag{3}$$

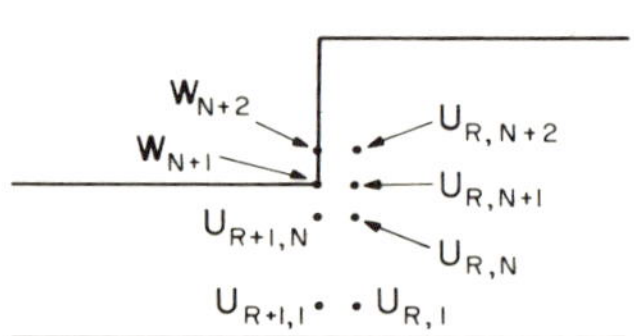

FIGURE 2

Assuming that $\tilde{u}_{R-1}$ has been determined by some method, we wish to determine u_R by an equation analogous to (6.3.1), that is by

$$u_R = \tilde{A}_{R-1} u_{R-1} + \tilde{b}_{R-1}, \tag{4}$$

where of course $\tilde{A}_{R-1}$ and $\tilde{b}_{R-1}$ are of order M. Since A_R and b_R also have order M, the $\tilde{A}_{R-1}$ and $\tilde{b}_{R-1}$ calculated by (6.3.4), namely

$$\begin{aligned} \tilde{A}_{R-1} &= [2I+Q-A_R]^{-1}, \\ \tilde{b}_{R-1} &= A_{R-1}(b_R+r_R), \end{aligned} \tag{5}$$

must be the desired matrix and vector. In other words, once $\tilde{u}_{R-1}$ is determined completely, the evaluation of u_R does not depend on the shape of the boundary to the left.* What we must do is ascertain how to compute u_{R-1} and thus $\tilde{u}_{R-1}$.

We would like to calculate u_{R-1} by an equation similar to (6.3.1)

$$u_{R-1} = A_{R-2} u_{R-2} + b_{R-2}, \tag{6}$$

where A_{R-2} and b_{R-2} are of order N. We will assume that A_{R-2} and b_{R-2} can be computed by means of

$$\begin{aligned} A_{R-2} &= [2I+Q-A_{R-1}]^{-1}, \\ b_{R-2} &= A_{R-2}(b_{R-1}+r_{R-1}), \end{aligned} \tag{7}$$

where all quantities appearing must have order N. We must now relate A_{R-1} to $\tilde{A}_{R-1}$ and b_{R-1} to $\tilde{b}_{R-1}$.

We let $\tilde{u}_R$ be defined as

$$\tilde{u}_R = (u_{R1}, \ldots, u_{RN}), \tag{8}$$

and try to calculate $\tilde{u}_R$ from u_{R-1} by means of a relation of the form

$$\tilde{u}_R = A_{R-1} u_{R-1} + b_{R-1}. \tag{9}$$

Finally we require u_R and $\tilde{u}_R$ to agree where both are defined (i.e., u_{R1}, $u_{R2}, \ldots, u_{RN}$) for *any choice of* u_{R-1} *and* w (or $\tilde{u}_{R-1}$). This procedure fixes the relationships between $\tilde{A}_{R-1}$ and A_{R-1} and between $\tilde{b}_{R-1}$ and b_{R-1}. More precisely, if we partition $\tilde{A}_{R-1}$ and $\tilde{b}_{R-1}$ in the following fashion

$$\tilde{A}_{R-1} = \begin{bmatrix} A_{11} & A_{12} \\ A_{21} & A_{22} \end{bmatrix}, \qquad \tilde{b}_{R-1} = \begin{bmatrix} b_1 \\ b_2 \end{bmatrix}, \tag{10}$$

* This observation is merely a restatement of the principle of causality for linear systems.

where A_{11} and b_1 have order N, then if we carry out the operations in (4) and (9), we find

$$\begin{aligned} A_{R-1} &= A_{11}, \\ b_{R-1} &= A_{12}w + b_1. \end{aligned} \tag{11}$$

4. Example

We will solve the potential equation in the region shown in Fig. 3 where the region changes at $R = 5$. In this situation w is the vector

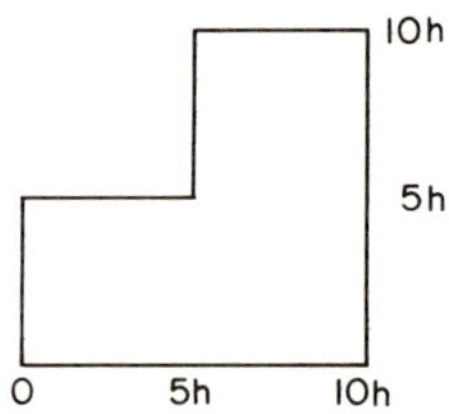

FIGURE 3

$$w = (u_{55}, u_{56}, u_{57}, u_{58}, u_{59}) \tag{1}$$

consisting of boundary values. We compute

$$\begin{aligned} A_{R-1} &= [2I+Q-A_R]^{-1}, \\ b_{R-1} &= A_{R-1}(b_R+r_R), \end{aligned} \tag{2}$$

starting with

$$A_9 = 0, \qquad b_9 = u_{10}, \tag{3}$$

where all the matrices and vectors are of order 9. After four iterations of (2) we have obtained

$$\begin{aligned} \tilde{A}_5 &= [2I+Q-A_6]^{-1}, \\ \tilde{b}_5 &= A_5(b_6+r_6). \end{aligned} \tag{4}$$

We compute A_5 and b_5 of order 4 by using (3.11). If

$$\begin{aligned} A_5 &= (a_{ij}), & i,j &= 1,\dots,4, \\ \tilde{A}_5 &= (\tilde{a}_{ij}), & i,j &= 1,\dots,9, \\ b_5 &= (b_i), & i &= 1,\dots,4, \\ b_5 &= (\tilde{b}_i), & i &= 1,\dots,9, \end{aligned} \tag{5}$$

we have

$$a_{ij} = \tilde{a}_{ij}, \qquad i, j = 1, \dots, 4,$$
$$b_i = \tilde{b}_i + \sum_{j=1}^{5} a_{i,j+4} w_{j+4}. \tag{6}$$

Now we proceed with (2) where all the matrices and vectors have order 4 until A_0 and b_0 are determined. The equation

$$u_{R+1} = A_R u_R + b_R \tag{7}$$

is solved starting with u_0 until u_5 is determined using the A_R and b_R of order four we found previously. We form the vector

$$\tilde{u}_5 = \begin{bmatrix} u_5 \\ w \end{bmatrix},$$

and proceed with (7) where now all quantities are of order nine.

5. Case II: Order $u_R <$ Order u_{R-1}

Again it is sufficient to consider the case shown in Fig. 4 for which

$$u_{R-1} = (u_{R-1,1}, u_{R-1,2}, \dots, u_{R-1,M}), \qquad M > N,$$
$$u_R = (u_{R1}, u_{R2}, \dots, u_{RN}), \tag{1}$$

and

$$w = (w_{N+1}, \dots, w_M) \tag{2}$$

is given by the boundary conditions. Proceeding as before, we set

$$\tilde{u}_R = \begin{bmatrix} u_R \\ w \end{bmatrix} = (u_{R1}, \dots, u_{RN}, w_{N+1}, \dots, w_M). \tag{3}$$

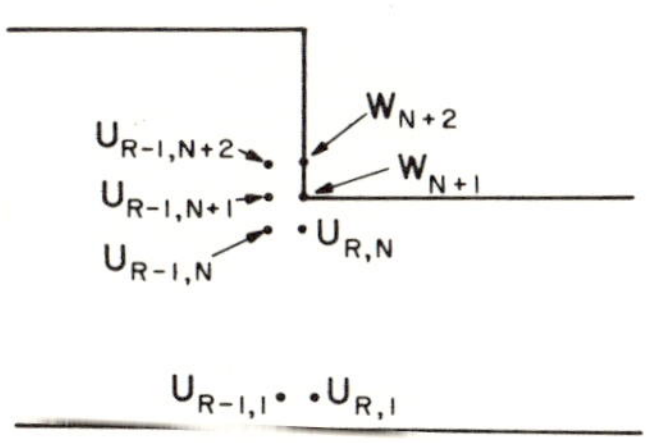

FIGURE 4

The matrix and vector of order M, A_R and b_R, are such that

$$A_{R-1} = [2I+Q-A_R]^{-1}, \quad b_{R-1} = A_{R-1}(b_R+r_R). \tag{4}$$

From A_{R-1} and b_{R-1} we would like to compute a matrix $\tilde{A}_{R-1}$ and a vector $\tilde{b}_{R-1}$ of order N such that $\tilde{u}_R$ can be computed as

$$\tilde{u}_R = \tilde{A}_{R-1} u_{R-1} + \tilde{b}_{R-1}. \tag{5}$$

Again we require that if $\tilde{u}_{R-1}$ is the vector

$$\tilde{u}_{R-1} = (u_{R-1,1}, \ldots, u_{R-1,M}), \tag{6}$$

then we can find u_R by

$$u_R = A_{R-1} \tilde{u}_{R-1} + b_{R-1}, \tag{7}$$

and we require (5) and (7) to agree for any u_{R-1}. We find immediately that

$$\tilde{A}_{R-1} = A_{R-1} \oplus 0, \qquad \tilde{b}_{R-1} = \begin{bmatrix} b_{R-1} \\ w \end{bmatrix}, \tag{8}$$

where $\oplus$ denotes the direct sum, i.e.,

$$\tilde{A}_{R-1} = \begin{bmatrix} A_{R-1} & 0 \\ 0 & 0 \end{bmatrix}. \tag{9}$$

We can look at this result as adjoining a new set of initial conditions as in (6.3.5.)

6. Example

To compute a solution of the potential equation in the region shown in Fig. 5, we proceed as follows. We calculate

$$A_{R-1} = [2I+Q-A_R]^{-1}, \quad b_{R-1} = A_{R-1}(b_R+r_R), \tag{1}$$

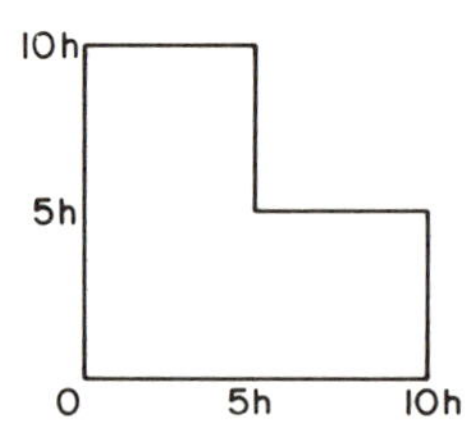

FIGURE 5

starting with

$$A_9 = 0, \qquad b_9 = u_{10}, \tag{2}$$

where all the matrices and vectors are of order four, until

$$\tilde{A}_5 = [2I + Q - A_6]^{-1}, \qquad \tilde{b}_5 = \tilde{A}_5(b_6 + r_6) \tag{3}$$

are determined. The vector

$$w = (u_{55}, u_{56}, u_{57}, u_{58}, u_{59}) \tag{4}$$

is given by the boundary conditions. The matrix A_5 and vector b_5 of order nine are formed by (5.8) or if

$$\begin{aligned} \tilde{A}_5 &= (\tilde{a}_{ij}), & i,j &= 1,\ldots,5, \\ A_5 &= (a_{ij}), & i,j &= 1,\ldots,9, \\ \tilde{b}_5 &= (\tilde{b}_i), & i &= 1,\ldots,4, \\ b_5 &= (b_i), & i &= 1,\ldots,9, \end{aligned} \tag{5}$$

then

$$\begin{aligned} a_{ij} &= \begin{cases} \tilde{a}_{ij}, & i,j = 1,\ldots,4, \\ 0, & \text{otherwise}, \end{cases} \\ b_i &= \begin{cases} \tilde{b}_i, & i = 1,\ldots,4, \\ w_i, & i = 5,\ldots,9. \end{cases} \end{aligned} \tag{6}$$

We proceed with (1) until A_0 and b_0, of order nine, are computed. The equation

$$u_{R+1} = A_R u_R + b_R \tag{7}$$

is solved starting with u_0 until

$$\tilde{u}_5 = \begin{bmatrix} u_5 \\ w \end{bmatrix} = A_4 u_4 + b_4$$

is found. Finally (7) is continued starting with u_5, where all the quantities are of order four.

7. Nonsingularity and Stability

In Chapter 4 we showed that a sufficient condition for both the nonsingularity of the matrices which must be inverted, and the computational stability of the method, was

$$0 < A_R < 1. \tag{1}$$

We will show here that this result still holds when the orders of u_R and u_{R-1} (and consequently A_R and A_{R-1}) are different.

For the case in which the order of u_{R-1} is greater than the order of u_R we have the following result. The inversion of the matrix $[2I+Q-A_R]$ yields a matrix $\tilde{A}_{R-1}$ which satisfies (1). The construction of A_{R-1} by means of (5.8) leaves the nonzero characteristic values of $\tilde{A}_{R-1}$ unchanged, and adds $N-M$ zero characteristic values. The matrix $[2I+Q-A_{R-1}]$ is clearly positive definite, and since

$$0 < A_{R-1} < 1, \tag{2}$$

we see that A_{R-2} must satisfy (1) as must all the subsequent matrices.

When the order of u_{R-1} is less than the order of u_R, we can apply the Sturmian separation Theorem to the operation required in (3.11). We see that (1) again must hold.

8. Removal of Restrictions

We are now in a position to discuss how the restrictions we made on the type of allowable region might be removed. We will not, however, discuss the validity of the discrete approximation of an irregular region since we have agreed previously that this topic is beyond the scope of this book.

For a given region it is clearly always possible to pick a grid spacing h such that all grid points are regular points except those next to boundary points. In this case we have to replace the finite difference approximations for points next to boundary points by more complicated expressions. For instance if $u(x-ah, y)$ is a boundary point then the second partial derivative at $u(x, y)$ will take the form

$$\frac{\partial^2 u}{\partial x^2}(x,y) \cong \frac{2}{h^2}\left[\frac{u(x+h,y)}{(1+a)} - \frac{u(x,y)}{a} + \frac{u(x-ah,y)}{a(1+a)}\right], \tag{1}$$

where

$$0 < a \leqslant 1. \tag{2}$$

In terms of our method this will mean slight changes in the definitions of Q and r_R and possibly some modifications of the equations for $R = 0$ and $R = N$.

If the region has a hole in it, we can handle this case by treating the parts of the region above and below the hole separately. For instance, if the regions is as in Fig. 6, then for $R_1 < R < R_2$ we have two A_R and two

FIGURE 6

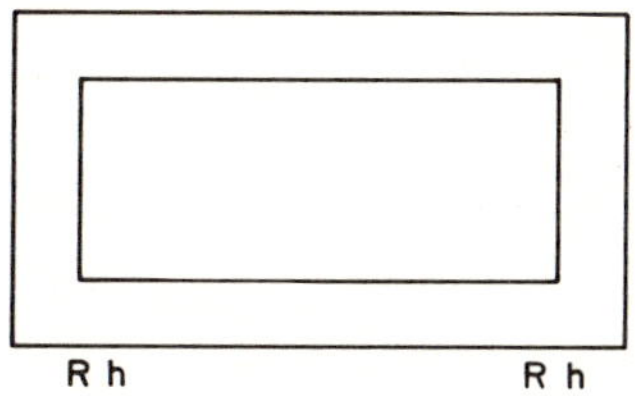

b_R, one set for the part of the region above the hole and the other for the part of the region below the hole. For $R_1 < R < R_2$ we carry out two independent sets of computation at each stage of the procedure. The results of Section 3 and Section 5 can easily be extended to tell us what to do at $R = R_1$ and $R = R_2$.

9. Examples

The potential equation was solved over the regions shown in Figs. 7–9. The boundary conditions were as shown in the figures. The results for the upper left hand quadrant of Fig. 8 are shown in Table I.

FIGURE 7

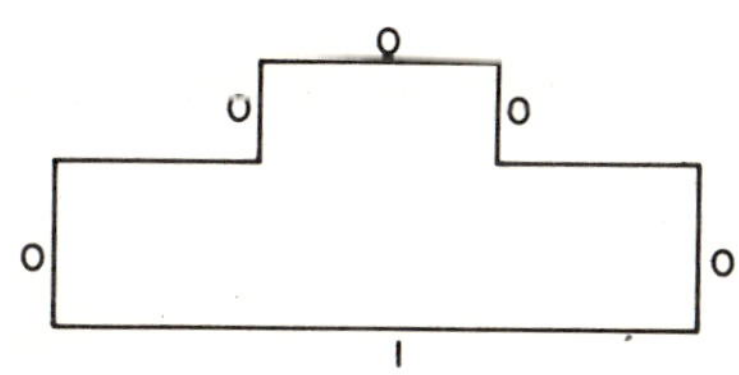

FIGURE 8

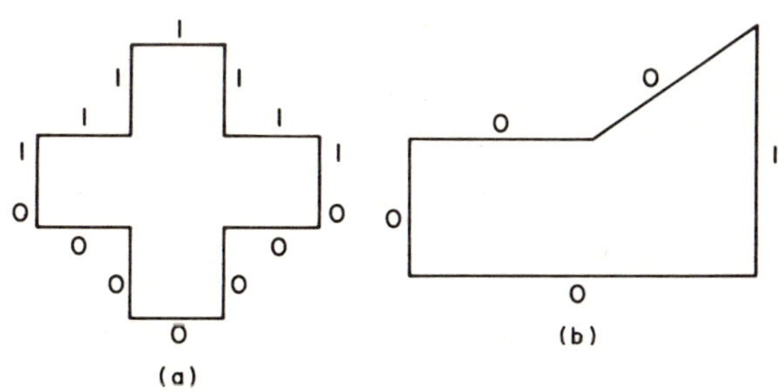

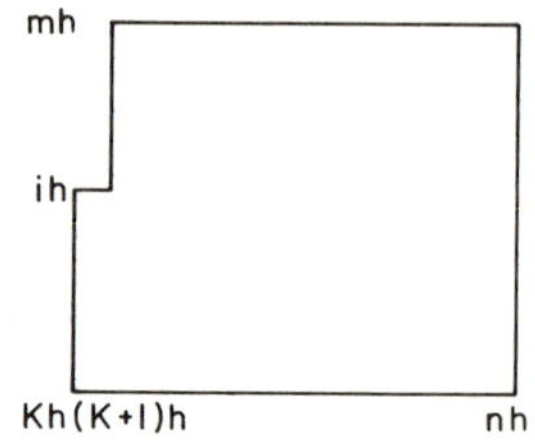

FIGURE 9

TABLE 1

				1.0	1.0	1.0	1.0	1.0
				1.0	0.9891	0.9803	0.9746	0.9727
				1.0	0.9662	0.9573	0.9455	0.9416
				1.0	0.9584	0.9272	0.9086	0.9025
				1.0	0.9303	0.8844	0.8592	0.8513
1.0	1.0	1.0	1.0	1.0	0.8786	0.8206	0.7927	0.7842
1.0	0.9104	0.8631	0.8379	0.8143	0.7635	0.7269	0.7067	0.7002
1.0	0.7786	0.7040	0.6744	0.6557	0.6340	0.6169	0.6068	0.6034
0.5	0.5000	0.5000	0.5000	0.5000	0.5000	0.5000	0.5000	0.5000

The important point to note is that all three problems were solved using *the same computer program*. The only modification necessary was that involved in changing a simple subroutine which describes the nature of the boundary.

10. General Linear Equations

Suppose we want to solve the equation

$$u_{xx} + u_{yy} - g(x,y)u + \varphi(x,y) = 0 \tag{1}$$

over the rectangle

$$0 \leqslant x \leqslant a, \qquad 0 \leqslant y \leqslant b, \tag{2}$$

where $u(x, y)$ is given on all sides of the rectangle. We use the same discretization as we have used so far. Thus, Q and r_R are the same and we define g_{ij} and φ_{ij} as in Chapter 5,

$$g_{ij} = g(ih, jh), \qquad \varphi_{ij} = \varphi(ih, jh). \tag{3}$$

The finite difference version of (1) is then

$$u_{i+1,j} + u_{i,j+1} - 4u_{ij} + u_{i-1,j} + u_{i,j-1} - h^2 g_{ij} u_{ij} + h^2 \varphi_{ij} = 0. \tag{4}$$

If we define matrices G_R and vectors φ_R as follows,

$$G_R = h^2 \operatorname{diag}(g_{R1}, g_{R2}, \ldots, g_{R,M-1}), \qquad \varphi_R = h^2 [\varphi_{Rj}], \tag{5}$$

the matrix–vector form of (4) is

$$u_{R+1} - 2u_R + u_{R-1} - Qu_R + r_R - G_R u_R + \varphi_R = 0, \tag{6}$$

with u_0 and u_N are given by the boundary conditions.

We look for solutions to (6) of the form

$$u_{R+1} = A_R u_R + b_R, \tag{7}$$

and we find

$$\begin{aligned} A_{R-1} &= [2I+Q+G_R-A_R]^{-1}, \\ b_{R-1} &= A_{R-1}(b_R+r_R+\varphi_R), \end{aligned} \tag{8}$$

with

$$A_{N-1} = 0, \qquad b_{N-1} = u_N. \tag{9}$$

A sufficient condition for stability and nonsingularity is as in Chapter 3,

$$G_R \geqslant 0. \tag{10}$$

Using this technique we can easily consider equations as general as

$$a_1 \frac{\partial^2 u}{\partial x^2} + a_2 \frac{\partial^2 u}{\partial y^2} + a_3 \frac{\partial^2 u}{\partial x\, \partial y} + a_4 \frac{\partial u}{\partial x} + a_5 \frac{\partial u}{\partial y} + a_6 u = \varphi(x,y), \tag{11}$$

where the a_i are functions of x and y.

11. Other Boundary Conditions

We can best illustrate how problems other than the Dirichlet problem can be handled by an example. Suppose we want to solve the potential equation

$$u_{xx} + u_{yy} = 0, \tag{1}$$

on the rectangle

$$0 \leqslant x \leqslant a, \qquad 0 \leqslant y \leqslant b, \tag{2}$$

with the boundary conditions

$$u(x,0) = f_1(x), \qquad u(x,b) = f_2(x), \tag{3}$$

and

$$\frac{\partial u}{\partial x}(0,y) = \frac{\partial u}{\partial x}(a,y) = 0. \tag{4}$$

The finite difference version of (1) is the same as before

$$u_{R+1} - 2u_R + u_{R-1} - Qu_R + r_R = 0. \tag{5}$$

Note, however, that u_0 and u_N are now not known.

Suppose we consider the vector u_{N+1} even though these points lie outside the region. The boundary condition

$$\frac{\partial u}{\partial x}(a,y) = 0, \tag{6}$$

can be approximated $O(h^2)$ by requiring, since $Nh = a$, that

$$u_{N+1} - u_{N-1} = 0. \tag{7}$$

Substituting (7) in (5) for $R = N$ we have

$$2u_{N-1} - 2u_N - Qu_N + r_N = 0. \tag{8}$$

However since we are looking for solutions of (5) of the form

$$u_{R+1} = A_R u_R + b_R, \tag{9}$$

then

$$u_N = A_{N-1} u_{N-1} + b_{N-1}. \tag{10}$$

We can now solve (8) for u_N in terms of u_{N-1} to find

$$u_N = [2I + Q]^{-1}(2u_{N-1} + r_N), \tag{11}$$

and thus by comparison with (10) we have

$$\begin{aligned} A_{N-1} &= 2[2I+Q]^{-1}, \\ b_{N-1} &= [2I+Q]^{-1} r_N. \end{aligned} \tag{12}$$

We now solve the usual equations

$$\begin{aligned} A_{R-1} &= [2I+Q-A_R]^{-1}, \\ b_{R-1} &= A_{R-1}(b_R + r_R), \end{aligned} \tag{13}$$

until A_0 and b_0 have been determined.

Finally we have to match the final boundary condition

$$\partial u(0,y)/\partial x = 0\,, \tag{14}$$

to find u_0 and u_1 so that we can use (9). If we look at u_{-1} which we assume exists, then an approximation of (14) is

$$u_{-1} - u_1 = 0\,. \tag{15}$$

Substituting (15) in (5) for $R = 0$ we find

$$2u_1 - 2u_0 - Qu_0 + r_0 = 0\,, \tag{16}$$

but by (9)

$$u_1 = A_0 u_0 + b_0\,. \tag{17}$$

We can solve these two equations for u_0,

$$u_0 = [2I + Q - 2A_0]^{-1}(2b_0 + r_0)\,, \tag{18}$$

and now we can proceed with (9) to calculate all the u_R.

If we are given boundary conditions such as (4) on the top and bottom boundaries, a little examination will show that we can handle this problem by modifying slightly our definitions of Q and r_R. In general, we can easily handle boundary conditions such as

$$au(x,y) + b[\partial u(x,y)/\partial x] = c \tag{19}$$

by these methods.

EXERCISE

1. Show that existence and stability hold for the above.

12. Three Dimensional Equations

Finally we would like to discuss some of the difficulties which arise when we try to solve three dimensional elliptic equations. To simplify the notation we will consider the potential equation

$$u_{xx} + u_{yy} + u_{zz} = 0 \tag{1}$$

on the cube

$$0 \leqslant x \leqslant a\,, \qquad 0 \leqslant y \leqslant a\,, \qquad 0 \leqslant z \leqslant a\,, \tag{2}$$

where $u(x, y, z)$ is assumed given on all sides of the cube. If we let

$$a = Nh, \tag{3}$$

and set

$$u_{ijk} = u(ih, jh, kh), \tag{4}$$

then the finite difference version of (1) is

$$u_{i+1,jk} + u_{i,j+1,k} + u_{ij,k+1} - 6u_{ijk} + u_{i-1,jk} + u_{i,j-1,k} + u_{i,j,k-1} = 0. \tag{5}$$

We will let the matrix U_R be defined as

$$U_R = (u_{Rjk}), \tag{6}$$

a surface of grid points for any R. The matrix Q will be defined as before, and the matrices S_R will be defined as

$$S_R = (s_{Rjk}), \qquad \text{where} \quad s_{Rjk} = \begin{cases} u_{Rj0}, & k = 1, \\ u_{RjN}, & k = N-1, \\ u_{R0,k}, & j = 1, \\ u_{RN,k}, & j = N-1, \\ 0, & \text{otherwise.} \end{cases} \tag{7}$$

and thus are known by the boundary conditions. Using this notation (5) can be written as the two-point boundary value problem

$$U_{R+1} - 2U_R + U_{R-1} - QU_R - U_R Q + S_R = 0, \tag{8}$$

with U_0 and U_N given by the boundary conditions. This is an extremely interesting equation which apparently cannot be handled by existing methods.

We might consider the following procedure. Let u_R and s_R be vectors of order N^2 derived by taking the elements of U_R and S_R by columns. Now (8) can be written as

$$u_{R+1} - 2u_R + u_{R-1} - [I \otimes Q + Q \otimes I]u_R + s_R = 0, \tag{9}$$

where $\otimes$ denotes the Kronecker product. We shall discuss Kronecker products in more detail in Chapter 8. Solutions to (9) can be found in the form

$$u_{R+1} = A_R u_R + b_R, \tag{10}$$

and the method can be shown always to yield a solution and to be computationally stable. These details are left as an exercise for the interested reader.

The difficulty is that the matrices that we must use are of order N^2. For reasonable N, e.g., $N = 10$, the problem is feasible for present day computers. For $N = 15$, the problem is currently too large. The N for feasible problems is increasing every year.

EXERCISES

1. Derive initial value problems for A_R and b_R in (10). Prove existence and stability.

2. Solve the three dimensional problem by dynimic programming.

13. The Biharmonic Equation

We will consider the following problem as an example of how our techniques apply to elliptic equations of fourth order. The static deflection of a simple supported square elastic plate under transverse loading is described by the biharmonic equation

$$u_{xxxx} + 2u_{xxyy} + u_{yyyy} = p\,, \tag{1}$$

where p is the loading. The boundary conditions are

$$u = 0\,, \qquad u_{xx} = 0\,, \tag{2}$$

on the edges $x = 0$ and $x = 1$, and

$$u = 0\,, \qquad u_{yy} = 0\,, \tag{3}$$

on the edges $y = 0$ and $y = 1$. As in Chapter 5, we begin by defining a new variable v through the equation

$$v = u_{xx} + u_{yy}\,. \tag{4}$$

Then (1) becomes

$$v_{xx} + v_{yy} = p\,, \tag{5}$$

and we now have a pair of coupled second order equations.

We can discretize (4) and (5) in the usual way. That is, we use the standard finite difference approximations for the second derivatives and convert the problem to matrix–vector form as we have done previously. Thus, (4) and

(5) become the coupled difference equations

$$\begin{aligned} v_{i+1} &= [2I + Q]\, v_i - v_{i-1} - p_i, \\ u_{i+1} &= [2I + Q]\, u_i - u_{i-1} + v_i, \end{aligned} \tag{6}$$

where Q is as defined previously and the vectors p_i are derived from the loading p. The appropriate boundray conditions for (6) are derived by discretizing (2) and (3).

Let us define vectors w_1 by

$$w_i = \begin{bmatrix} u_i \\ v_i \end{bmatrix} \tag{7}$$

so that (6) becomes

$$w_{i+1} = \begin{bmatrix} 2I + Q & I \\ 0 & 2I + Q \end{bmatrix} w_i - w_{i-1} - \begin{bmatrix} 0 \\ p_1 \end{bmatrix}. \tag{8}$$

This is a problem we can solve easily by either invariant imbedding or the Riccati transformation using the foregoing procedures.

EXERCISES

1. Derive appropriate boundary conditions for (8).

2. What is the Riccati equation associated with (8)? Prove existence and stability.

3. Use continuous methods to reduce (4) and (5) to the equations of Chapter 6, Section 13.

14. Invariant Imbedding and Difference Equations

The mathematical description of invariant imbedding has been somewhat clouded by the fact that the original invariant imbedding derivations were almost always arrived at using physical reasoning. The purpose of this section will be to explore the relation between invariant imbedding and linearity in a rigorous manner.

We will consider the matrix–vector system of difference equations

$$\begin{bmatrix} u(i+1) \\ v(i+1) \end{bmatrix} = \begin{bmatrix} A(i) & B(i) \\ C(i) & D(i) \end{bmatrix} \begin{bmatrix} u(i) \\ v(i) \end{bmatrix} + \begin{bmatrix} e(i) \\ f(i) \end{bmatrix}, \tag{1}$$

with the boundary conditions

$$u_M = a, \qquad v_N = b, \qquad N > M. \tag{2}$$

We will let $u(i)$ be a k-dimensional vector and $v(i)$ be an l-dimensional vector. Then $A(i)$, $B(i)$, $C(i)$, $D(i)$, $e(i)$ and $f(i)$ are $k\times k$, $l\times k$, $k\times l$, $l\times l$, $k\times l$, and $l\times l$ matrices respectively. We write $u(i)$ and $v(i)$ as

$$u(i) = u(i,M,N,a,b), \qquad v(i) = v(i,M,N,a,b) \tag{3}$$

to emphasize the dependence of $u(i)$ and $v(i)$ on the interval length and the boundary conditions.

By linearity we have

$$\begin{aligned} u(i,M,N,A,B) &= W(i,M,N)a + R(i,M,N)b + r(i,M,N), \\ v(i,M,N,a,b) &= S(i,M,N)a + T(i,M,N)b + s(i,M,N), \end{aligned} \tag{4}$$

or

$$\begin{bmatrix} u(i) \\ v(i) \end{bmatrix} = \begin{bmatrix} W(i,M,N) & R(i,M,N) \\ S(i,M,N) & T(i,M,N) \end{bmatrix} \begin{bmatrix} a \\ b \end{bmatrix} + \begin{bmatrix} r(i,M,N) \\ s(i,m,n) \end{bmatrix}. \tag{5}$$

The functions W, R, T and S are functions of only the position and the interval. We will show that we can find initial value problems which will yield these functions. Applying the boundary conditions to (5) we find

$$\begin{aligned} &W(M,M,N) = I, \qquad R(M,M,M) = 0, \qquad r(M,M,N) = 0, \\ &S(N,M,N) = 0, \qquad T(N,M,N) = I, \qquad s(N,M,N) = 0. \end{aligned} \tag{6}$$

We will consider the point

$$\bar{a} = u(M+1,M,N,a,b). \tag{7}$$

By causality

$$u(i,M,N,a,b) = u(i,M+1,N,\bar{a},b), \tag{8}$$

$$v(i,M,N,a,b) = v(i,M+1,N,\bar{a},b). \tag{9}$$

In other words the values $u(i)$ and $v(i)$ are the same whether we start the process with a at M or $u(M+1)$ at $M+1$. Then (5) becomes

$$\begin{aligned} &\begin{bmatrix} W(i,M,N) & R(i,M,N) \\ S(i,M,N) & T(i,M,N) \end{bmatrix} \begin{bmatrix} a \\ b \end{bmatrix} + \begin{bmatrix} r(i,M,N) \\ s(i,M,N) \end{bmatrix} \\ &= \begin{bmatrix} W(i,M+1,N) & R(i,M+1,N) \\ S(i,M+1,N) & T(i,M+1,N) \end{bmatrix} \begin{bmatrix} a \\ b \end{bmatrix} + \begin{bmatrix} r(i,M+1,N) \\ s(i,M+1,N) \end{bmatrix}, \end{aligned} \tag{10}$$

$$\begin{bmatrix} \bar{a} \\ b \end{bmatrix} = \begin{bmatrix} W(M+1,M,N) & R(M+1,M,N) \\ 0 & I \end{bmatrix} \begin{bmatrix} a \\ b \end{bmatrix} + \begin{bmatrix} r(M+1,M,N) \\ 0 \end{bmatrix}. \tag{11}$$

Substituting this expression in (10) and letting $i = M+1$, we get using (6)

$$\begin{aligned} &\begin{bmatrix} W(M+1,M,N) & R(M+1,M,N) \\ S(M+1,M,N) & T(M+1,M,N) \end{bmatrix} \begin{bmatrix} a \\ b \end{bmatrix} + \begin{bmatrix} r(M+1,M,N) \\ s(M+1,M,N) \end{bmatrix} \\ &\quad = \begin{bmatrix} I & 0 \\ S(M+1,M+1,N) & T(M+1,M+1,N) \end{bmatrix} \\ &\quad \times \left\{ \begin{bmatrix} W(M+1,M,N) & R(M+1,M,N) \\ 0 & I \end{bmatrix} \begin{bmatrix} a \\ b \end{bmatrix} \right. \\ &\quad \left. + \begin{bmatrix} r(M+1,M,N) \\ 0 \end{bmatrix} \right\} + \begin{bmatrix} 0 \\ s(M+1,M+1,N \end{bmatrix}. \end{aligned} \tag{12}$$

Using (5) in the original equations and letting $i = M$, we see

$$\begin{aligned} &\begin{bmatrix} W(M+1,M,N) & R(M+1,M,N) \\ S(M+1,M,N) & T(M+1,M,N) \end{bmatrix} \begin{bmatrix} a \\ b \end{bmatrix} + \begin{bmatrix} r(M+1,M,N) \\ s(M+1,M,N) \end{bmatrix} \\ &= \begin{bmatrix} A(M) & B(M) \\ C(M) & D(M) \end{bmatrix} \left\{ \begin{bmatrix} I & 0 \\ S(M,M,N) & T(M,M,N) \end{bmatrix} \begin{bmatrix} a \\ b \end{bmatrix} \right. \\ &\quad \left. + \begin{bmatrix} 0 \\ s(M,M,N) \end{bmatrix} \right\} + \begin{bmatrix} e(M) \\ f(M) \end{bmatrix}, \end{aligned} \tag{13}$$

and by comparison with (12) we find

$$\begin{aligned} &\begin{bmatrix} I & 0 \\ \bar{S}(M+1) & \bar{T}(M+1) \end{bmatrix} \begin{bmatrix} W(M+1,M,N) & R(M+1,M,N) \\ 0 & 0 \end{bmatrix} \\ &\qquad \times \begin{bmatrix} a \\ b \end{bmatrix} \begin{bmatrix} r(M+1,M,N) \\ 0 \end{bmatrix} + \begin{bmatrix} 0 \\ \bar{s}(M+1) \end{bmatrix} \\ &\quad = \begin{bmatrix} A(M) & B(M) \\ C(M) & D(M) \end{bmatrix} \begin{bmatrix} I & 0 \\ \bar{R}(M) & \bar{T}(M) \end{bmatrix} \begin{bmatrix} a \\ b \end{bmatrix} + \begin{bmatrix} 0 \\ s(M) \end{bmatrix} \\ &\qquad + \begin{bmatrix} e(M) \\ f(M) \end{bmatrix}, \end{aligned} \tag{14}$$

where we have defined

$$\begin{aligned} \bar{S}(M) &= S(M, M, N), \\ \bar{T}(M) &= T(M, M, N), \\ \bar{s}(M) &= s(M, M, N). \end{aligned} \tag{15}$$

The first equation of (1) can be rewritten by applying (4) as

$$\begin{aligned} &u(M+1, M, N, a, b) \\ &\quad = A(M)\,u(M, M, N, a, b) + B(M)\,v(M, M, N, a, b) + e(M) \\ &\quad = A(M)\,a + B(M)\,[\bar{S}(M)\,a + \bar{T}(M)\,b + \bar{s}(M)]\,e(M). \end{aligned} \tag{16}$$

However, also by (4)

$$\begin{aligned} &u(M+1, M, N, a, b) \\ &\quad = W(M+1, M, N)\,a + R(M+1, M, N)\,b + r(M+1, M, N), \end{aligned} \tag{17}$$

and thus by equating coefficients in (16) and (17) we find

$$\begin{aligned} W(M+1, M, N) &= A(M) + B(M)\,\bar{R}(M), \\ R(M+1, M, N) &= B(M)\,\bar{T}(M), \\ r(M+1, M, N) &= B(M)\,\bar{s}(M) + e(M). \end{aligned} \tag{18}$$

We can now substitute these equations in (14) and reduce the system to the following one involving only the functions $\bar{T}(M)$, $\bar{S}(M)$ and $\bar{s}(M)$,

$$\begin{aligned} &\begin{bmatrix} A(M) & B(M) \\ C(M) & D(M) \end{bmatrix} \left\{ \begin{bmatrix} I & 0 \\ \bar{S}(M) & \bar{T}(M) \end{bmatrix} \begin{bmatrix} a \\ b \end{bmatrix} + \begin{bmatrix} 0 \\ \bar{s}(M) \end{bmatrix} \right\} + \begin{bmatrix} e(M) \\ f(M) \end{bmatrix} \\ &= \begin{bmatrix} I & 0 \\ \bar{S}(M+1) & \bar{T}(M+1) \end{bmatrix} \left\{ \begin{bmatrix} A(M) + B(M)\,\bar{S}(M) & B(M)\,\bar{T}(M) \\ 0 & I \end{bmatrix} \right. \\ &\quad \times \left. \begin{bmatrix} a \\ b \end{bmatrix} + \begin{bmatrix} B(M)\,\bar{s}(M) + e(M) \\ 0 \end{bmatrix} \right\} + \begin{bmatrix} 0 \\ \bar{s}(M+1) \end{bmatrix}. \end{aligned} \tag{19}$$

Finally by equating coefficients of a and b in this expression we find equations for $\bar{S}(M)$, $\bar{T}(M)$ and $\bar{s}(M)$

$$\begin{aligned} \bar{S}(M) &= (D(M) - \bar{S}(M+1)\,B(M))^{-1}\,(\bar{S}(M+1)A(M) - C(M)), \\ \bar{T}(M) &= (D(M) - \bar{S}(M+1)\,B(M))^{-1}\,\bar{T}(M+1), \\ \bar{s}(M) &= (D(M) - \bar{S}(M+1)\,B(M))^{-1} \\ &\qquad \times (\bar{s}(M+1) + \bar{S}(M+1)\,e(M) - f(M)). \end{aligned} \tag{20}$$

The initial conditions are given by (6) as

$$\bar{S}(N) = 0, \qquad \bar{T}(N) = I, \qquad \bar{s}(N) = 0. \tag{21}$$

We can now proceed in the same manner and investigate the functions $R(i, M, N)$, $W(i, M, N)$ and $s(i, M, N)$. We consider the point

$$\bar{b} = v(N+1, M, N, a, b). \tag{22}$$

If we define $\bar{R}(N)$, $\bar{W}(N)$ and $\bar{s}(N)$ as

$$\bar{R}(N) = R(N, M, N), \qquad W(N) = \bar{W}(N, M, N), \qquad \bar{s}(N) = s(N, M, N), \tag{23}$$

we find

$$\begin{aligned} \bar{R}(N+1) &= (A(N)\bar{R}(N) + B(N))(C(N)\bar{R}(N) + D(N))^{-1}, \\ \bar{W}(N+1) &= A(N)\bar{W}(N) - \bar{R}(N+1), \\ \bar{s}(N+1) &= A(N)\bar{s}(N) + e(N) - \bar{R}(N+1)\, CC(N)\bar{s}(N) + f(N)), \end{aligned} \tag{24}$$

with the initial conditions

$$R(M) = 0, \qquad W(M) = I, \qquad s(M) = 0. \tag{25}$$

Although we can now construct all the functions in (4), these are not necessary to solve the original problem. All we need are either the functions defined by (20) or those defined by (25). For, by causality, we can write

$$\begin{bmatrix} u(i) \\ v(i) \end{bmatrix} = \begin{bmatrix} I & 0 \\ \bar{S}(i) & \bar{T}(i) \end{bmatrix} \begin{bmatrix} u(i) \\ b \end{bmatrix} + \begin{bmatrix} 0 \\ \bar{s}(i) \end{bmatrix} \tag{26}$$

and substitute this expression into (1) to obtain

$$\begin{bmatrix} u(i+1) \\ v(i+1) \end{bmatrix} = \begin{bmatrix} A(i) & B(i) \\ C(i) & D(i) \end{bmatrix} \left\{ \begin{bmatrix} I & 0 \\ \bar{S}(i) & \bar{T}(i) \end{bmatrix} \begin{bmatrix} u(i) \\ b \end{bmatrix} + \begin{bmatrix} 0 \\ \bar{s}(i) \end{bmatrix} \right\} + \begin{bmatrix} e(i) \\ f(i) \end{bmatrix}, \tag{27}$$

which is an initial value problem starting with

$$u(M) = a, \tag{28}$$

once (20) has been solved. We can also derive the initial value problem

$$\begin{bmatrix} u(i+1) \\ v(i+1) \end{bmatrix} = \begin{bmatrix} A(i) & B(i) \\ C(i) & D(i) \end{bmatrix} \left\{ \begin{bmatrix} \overline{W}(i) & \overline{R}(i) \\ 0 & I \end{bmatrix} \begin{bmatrix} a \\ v(i) \end{bmatrix} + \begin{bmatrix} s(i) \\ 0 \end{bmatrix} \right\} + \begin{bmatrix} e(i) \\ f(i) \end{bmatrix} \tag{29}$$

in a similar manner.

EXERCISES

1. Fix i and vary N and M in (3).
2. What happens for coupled ordinary differential equations?

15. A Second Approach

Let us now attempt a second building block technique which will allow us to solve partial differential equations over general regions using only scalar computations. Once again our example will be the potential equation with the values on the boundary given.

Suppose we consider the solution of the potential equation over the region shown in Fig. 10. We assume that the discretization has been done as in Fig. 11 and that the left boundary values are denoted by the quantities $\{v_k\}$, i.e.,

$$\begin{aligned} v_j &= u_{jk}, & j &= 1, \ldots, l-1, \\ v_j &= u_{j,k+1}, & j &= l+1, \ldots, m-1. \end{aligned} \tag{1}$$

Let us denote by $f_{lk}(v_1, \ldots, v_{m-1})$ the value of the discrete functional over the mesh in Fig. 11 where, as usual, the left boundary values are considered

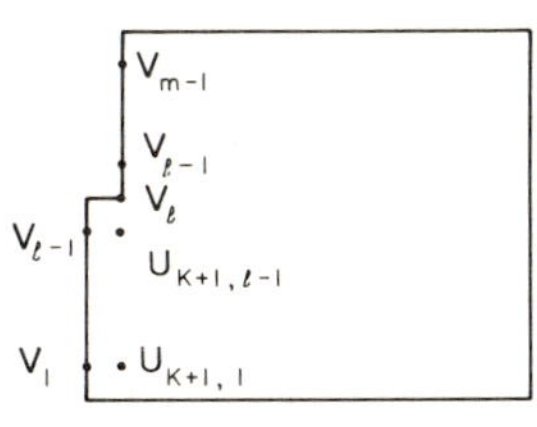

FIGURE 10

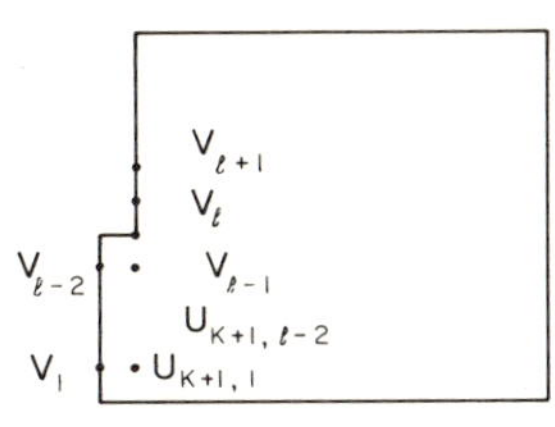

FIGURE 11

as parameters. Thus, we have by the same procedure as in Chapter 5

$$f_{lk}(v_1, \ldots, v_{m-1}) = \min_{\{u_{ij}\}} \left[\sum_{j=1}^{l} (u_{k+1,j} - u_{k+1,j-1})^2 + \sum_{j=1}^{l-1} (u_{k+1,j} - u_{k,j})^2 + \sum_{l=k+2}^{n} \left[\sum_{j=1}^{m} (u_{ij} - u_{i,j-1})^2 + \sum_{j=1}^{m-1} (u_{ij} - u_{i-1,j})^2 \right] \right], \quad l = 2, 3, \ldots, m, \tag{2}$$

where the $\{v_i\}$ are as in (1) and the minimization is carried out over the interior points. Expanding the above expression we can write

$$f_{lk}(v_1, \ldots, v_{m-1}) = \min_{u_{k+1,l-1}} \Bigg[(v_l - u_{k+1,l-1})^2 + (v_{l-1} - u_{k+1,l-1}) + \min_{\{u_{ij}\}} \left[\sum_{j=1}^{l-1} (u_{k+1,j} - u_{k+1,j-1})^2 + \sum_{j=1}^{l-2} (u_{k+1,j} - u_{kj})^2 + \sum_{i=k+2}^{n} \left[\sum_{j=1}^{m} (u_{ij} - u_{ij-1})^2 + \sum_{j=1}^{m-1} (u_{ij} - u_{i-1,j})^2 \right] \right] \Bigg], \tag{3}$$

or in view of (2), letting $\omega = u_{k+1,l+1}$

$$f_{lk}(v_1, \ldots, v_{m-1}) = \min_{\omega} [(v_l - \omega)^2 + (v_{l-1} - \omega)^2 + f_{l-1,k}(v_1, \ldots, v_{l-2}, \omega, v_l, \ldots, v_{m-1})]. \tag{4}$$

This important result is another restatement of the principle of optimality. We have succeeded in relating the problem over the region of Fig. 11 with that over the region of Fig. 13. We should expect a result of this form since when we go from the region of Fig. 12 to that of Fig. 13, we add two additional terms to our discrete functional and must minimize over one additional interior point.

Solving (4) numerically, a topic we shall return to shortly, will enable us to proceed from the region of Fig. 13 to that of Fig. 14 by a sequence of *scalar* minimizations. However, we would also like to find f_{1k} by scalar operations. To accomplish this, we start by defining $f_{0,k}(v_1, \ldots, v_{m-1})$ as

$$f_{0,k}(v_1, \ldots, v_{m-1}) = f_{m,k+1}(v_1, \ldots, v_{m-1}), \tag{5}$$

an obvious choice upon consideration of Fig. 14.

Now consider the regions in Fig. 15 and Fig. 16. In both regions the minimizations are carried out over the same interior points. If we let

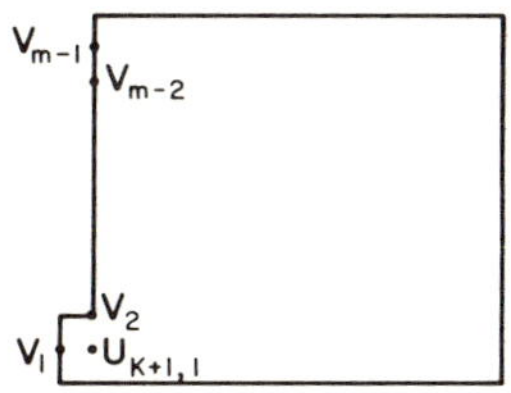

FIGURE 12

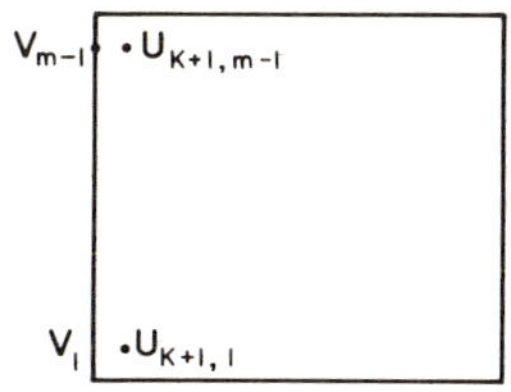

FIGURE 13

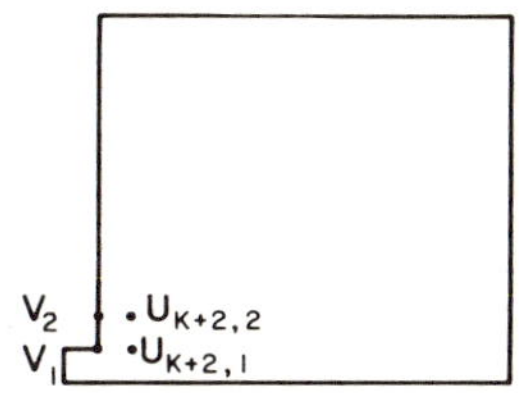

FIGURE 14

$f_{1k}(v_1, \ldots, v_{m-1})$ denote the minimum of the discrete function over the region of Fig. 15, and using (5), let $f_{0k}(v_1, \ldots, v_{m-1})$ be the minimum over the region of Fig. 16, we have

$$f_{1,k}(v_1, \ldots, v_{m-1}) = (v_1 - u_{k+1,0})^2 + f_{0,k}(v_1, \ldots, v_{m-1}). \tag{6}$$

To determine $f_{m,n-1}(v_1, \ldots, v_{m-1})$ we consider the region of Fig. 17. Since

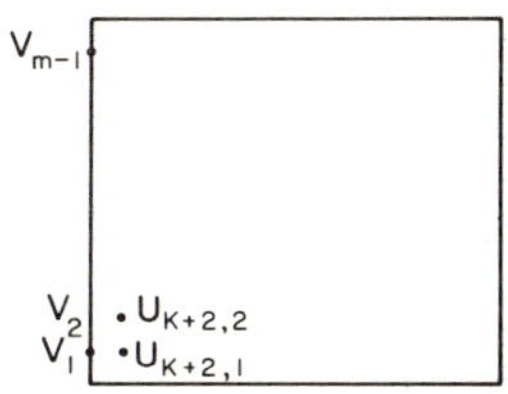

FIGURE 15

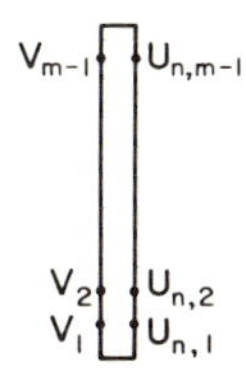

FIGURE 16

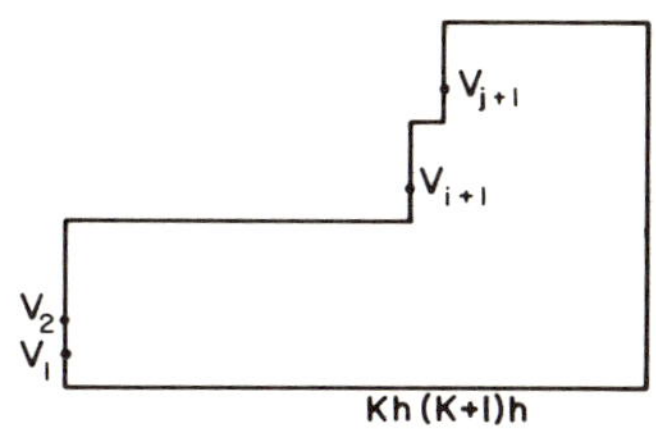

FIGURE 17

there are no interior points we have

$$f_{m,n-1}(v_1, \ldots, v_{m-1}) = \sum_{j=1}^{m} (u_{nj} - u_{n,j-1})^2 + \sum_{j=1}^{m-1} (v_j - u_{n,j})^2, \tag{7}$$

an expression involving only the $\{v_i\}$ and boundary values. Hence to solve the potential equation we need only find $f_{ik}(v_1, \ldots, v_m)$ for the proper i and n. We do this by solving the sequence of scalar problems

$$f_{m,n-1} = f_{0,n-2}, f_{1,n-2}, \ldots, f_{m-1,n-2},$$
$$f_{m,m-2} = f_{0,n-3}, f_{1,n-3}, \ldots, f_{i-1,k}, f_{ik}.$$

16. Matrix-Vector Equations

We will now turn to the computational aspects of our new method. Let us denote by v the $m-1$ dimensional vector

$$v = [v_i], \qquad i = 1, \ldots, m-1. \tag{1}$$

Once again we can establish inductively that $f_{lk}(v, \ldots, v_{m-1}) = f_{lk}(v)$ is quadratic in v,

$$f_{lk}(v) = (A^{(l,k)} v, v) - 2(b^{(l,k)}, v) + c^{(l,k)} . \tag{2}$$

In fact, using the notation of previous sections, we can easily verify that

$$f_{m,n-1}(v) = (v, v) - 2(u_n, v) + ([I + Q] u_n, u_n) - (2r_n, u_n) + s_n . \tag{3}$$

Hence

$$A^{(m,n-1)} = I, \qquad b^{(m,n-1)} = u_n . \tag{4}$$

Since as usual we are only interested in the $\{u_{ij}\}$, we will not need the values of $c^{(l,k)}$ and we will neglect all scalar terms. From (15.5) we have

$$A^{(0,k)} = A^{(m,k+1)}, \qquad b^{(0,k)} = b^{(m,k+1)} . \tag{5}$$

With no loss of generality we can take $A^{(l,k)}$ to be symmetric. Let us use the notation

$$A^{(l,k)} = (a_{ij}^{(l,k)}), \qquad b^{(l,k)} = (b_i^{(l,k)}) . \tag{6}$$

Then by (13.6) we have

$$\begin{aligned} a_{ij}^{(i,k)} &= \begin{cases} a_{11}^{(0,k)} + 1 & \text{if } i = j = 1, \\ a_{ij}^{(0,k)} & \text{otherwise,} \end{cases} \\ b_i^{(i,k)} &= \begin{cases} b_1^{(0,k)} + u_{k+1,0} & \text{if } i = 1, \\ b_i^{(0,k)} & \text{otherwise.} \end{cases} \end{aligned} \tag{7}$$

Let us write (15.4), using (2), as

$$f_{lk}(v) = \min_{\omega} [(v_l - \omega)^2 + (v_{l-1} - \omega)^2] + \left(A^{(l-1,k)} \begin{bmatrix} v_1 \\ \vdots \\ \omega \\ v_l \\ \vdots \\ v_m \end{bmatrix}, \begin{bmatrix} v_1 \\ \vdots \\ \omega \\ v_l \\ \vdots \\ v_m \end{bmatrix} \right) - 2 \left(b^{(l-1,k)}, \begin{bmatrix} v_1 \\ \vdots \\ \omega \\ v_l \\ \vdots \\ v_{m-1} \end{bmatrix} \right) + c^{(l-1,k)}. \tag{8}$$

We can differentiate this expression to find the minimizing ω,

$$\omega = v_l + \frac{v_{l-1} + b_l^{(l-1,k)} - \sum_{\substack{i=1 \\ i \neq l}}^{m-1} a_{li} v_i}{2 + a_{ll}^{(l-1,k)}} . \tag{9}$$

We shall now use this result in (8) to find $A^{(l,k)}$ and $b^{(l,k)}$. To simplify the notation, let us write

$$\begin{aligned} \omega - v_l &= (\alpha, v) + \beta, \\ \omega - v_{l-1} &= (\alpha', v) + \beta, \end{aligned} \tag{10}$$

where by (9)

$$\alpha_i = \begin{cases} \dfrac{-(1 + a_{ll}^{(l-1,k)})}{2 + a_{ll}^{(l-1,k)}}, & i = l, \\[2ex] \dfrac{1 - a_{l,l-1}}{2 + a_{ll}^{(l-1,k)}}, & i = l - 1, \\[2ex] \dfrac{-a_l}{2 + a_{ll}^{(l-1,k)}}, & \text{otherwise}, \end{cases}$$

$$\alpha_i' = \begin{cases} \dfrac{1}{2 + a_{ll}^{(l-1,k)}}, & i = l, \\[2ex] \dfrac{-(1 + a^{(l-1,k)} + a_{ll-1}^{(l-1,k)})}{2 + a_{ll}^{(l-1,k)}}, & i = l - 1, \\[2ex] \dfrac{-a_{li}}{2 + a_{ll}^{(l-1,k)}}, & \text{otherwise}, \end{cases} \tag{11}$$

$$\beta = \frac{b_l^{(l-1,k)}}{2 + a_{ll}^{(l-1,k)}}.$$

Then

$$\begin{aligned} f_{lk}(v) = {} & [(\alpha, v) + \beta]^2 + [(\alpha', v) + \beta]^2 + (A^{(l-i,k)}(v + \delta v), v + \delta v) \\ & - 2(b^{(l-1,k)}, v + \delta v) + c^{(l-1,k)}, \end{aligned} \tag{12}$$

where by (10)

$$\delta v = [\delta v_i],$$

$$\delta_{vi} = \begin{cases} (\alpha, v) + \beta, & i = l - 1, \\ 0 & \text{otherwise}. \end{cases} \tag{13}$$

Expanding (12) and writing out all terms in v we find

$$(\alpha, v)^2 = (Av, v), \tag{14}$$

where

$$A = (a_{ij}) = (\alpha_i \alpha_j). \tag{15}$$

Likewise

$$(\alpha', v)^2 = (A'^2 v, v), \tag{16}$$

and

$$\sum_{i=1}^{m-1} a_{l-1,j} v_i \sum_{j=1}^{m-1} \alpha_j v_j = (A'' v, v), \tag{17}$$

where

$$\begin{aligned} A' &= (a'_{jj}) = (\alpha'_i \alpha'_j), \\ A'' &= (a''_{ij}) = (a^{(l-1,k)}_{(l-1 j \alpha j)}). \end{aligned} \tag{18}$$

Thus, let

$$\begin{aligned} \bar{A}^{(l,k)} &= A^{(l-1k)} + A + A' + A'' \\ &= (a^{(l-1,k)}_{ij} + (a^{(l-1,k)}_{l-1i} + 2\alpha_i)\alpha_j + \alpha'_i \alpha'_j). \end{aligned} \tag{19}$$

Finally we determine $A^{(l,k)}$ from $\bar{A}^{(l,k)}$ by requiring this matrix to be symmetric, i.e.,

$$A^{(l,k)} = (a^{(l,k)}_{ij}) = \tfrac{1}{2}(\bar{a}^{(l,k)}_{ij} + \bar{a}^{(l,k)}_{ji}). \tag{20}$$

In a similar manner we find

$$\begin{aligned} b^{(l,k)} = [b^{(l,k)}_i] = [&-\beta(2\alpha_i + \alpha'_j) - a^{(l-1,k)}_{l-1,j} \\ &+ b^{(l-1,k)}_i + b^{(l-1,k)}_{l-1}\alpha_i]. \end{aligned} \tag{21}$$

Thus we can compute and store the matrices $A^{(l,k)}$ and vectors $b^{(l,k)}$ in a straightforward manner. Then we use (9) with $\omega = u_{k,l-1}$, to construct the solution.

17. General Regions

It is clear that using the ideas of the last two sections we can treat general two dimensional regions by a sequence of scalar minimizations. Our main difficulty is notational. For instance, for the region of Fig. 18 we might adopt the notation $f_{ijk}(v_1, \ldots, v_m)$ to denote the minimum value of the discrete functional over this region. Proceding as before we would be led to a functional equation of the form

$$\begin{aligned} f_{ijk}(v_1, \ldots, v_{m-1}) = \min_w [&(v_j - w)^2 + (v_{j-1} - w)^2 \\ &+ f_{i,j-1,k}(v_1, \ldots, v_{j-2}, w, v_j, \ldots, v_{m-1})] \end{aligned} \tag{1}$$

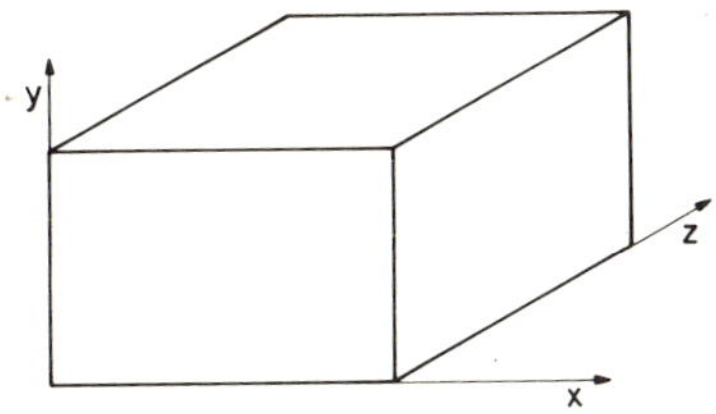

FIGURE 18

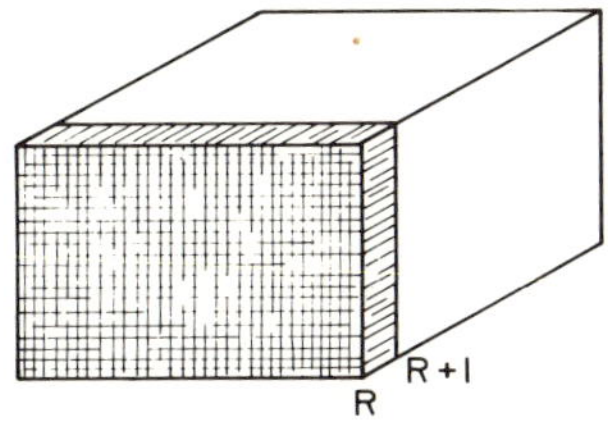

FIGURE 19

for $j > i$, and other relations similar to those of the last two sections for values of i, j, and k along the boundaries.

However the most interesting and important application of these results is to the solution of partial differential equations over three (or more) dimensional regions. Let us consider, for example, the region of Fig. 19, where

$$a = Nh, \qquad b = Mh, \qquad c = Lh. \tag{2}$$

If we let u denote the solution of the potential equation

$$u_{xx} + u_{yy} + u_{zz} = 0, \tag{3}$$

over this region with the value of u specified on the boundary, then we can replace (3) by the variational problem

$$\min J(u) = \min \int\int_R\int (u_x^2 + u_y^2 + u_z^2)\, dR, \tag{4}$$

subject to the same boundary conditions as (3). We now discretize (4) and consider only the values of u at grid points. Let $\{u_{jk}\}$ denote these grid points, i.e.,

$$u_{ijk} = u(ih, jh, kh). \tag{5}$$

We now apply dynamic programming to the discrete version of (4) which is

$$\min_{\{u_{ijk}\}} \sum_{i=1}^{N} \sum_{j=1}^{M} \sum_{k=1}^{L} [(u_{ijk} - u_{i-1,j,k})^2 + (u_{ijk} - u_{i,j-1,k})^2 + (u_{ijk} - u_{ij,k-1})^2]. \tag{6}$$

We define $N-1 > M-1$ matrices U_l in an analogous fashion to the vectors u_R of two-dimensional case, i.e.,

$$U_l = [u_{ijl}]. \tag{7}$$

Applying the principle of optimality we obtain a functional equation of the form

$$f_l(V) = \min_{U_l} [(QU_l, U_l) - (2R_l, U_R) + S_L + f_{l+1}(U_R)], \tag{8}$$

where the inner product of two matrices is defined as

$$(A, B) = \sum_{i=1}^{n} \sum_{j=1}^{n} a_{ij} b_{ij} = \operatorname{tr}(AB). \tag{9}$$

We can now use the fact that once again $f_l(V)$ must be quadratic in V,

$$f_l(V) = \langle A_l V, V \rangle - 2\langle B_l, V \rangle + C_l. \tag{10}$$

Hence we can easily derive matrix recurrence relations for the matrices A and B. Unfortunately the matrices we must invert will be of order $(N-1)(M-1)$ since our imbedding shows how the solution changes as a whole surface of grid points is removed as in Fig. 20. Thus for even moderate values of M and N the matrices we must invert can be too large to handle in most computers. We will see in later chapters how this difficulty also arises in the invariant imbedding approach.

Suppose now we consider the region of Fig. 21. We can let $f_{lk}(v_1, \ldots, v_m)$ now denote the minimum of the discrete functional over this region where

$$v_i = u_{i,k+1,l} \tag{11}$$

which are values along the boundary. If we proceed as with the two-dimensional case we can remove the $M-1$ points as in Fig. 22 and relate $f_{lk}(v)$ to $f_{lk+1}(v)$. Now since we are removing $M-1$ points at a time we will have matrices of order $M-1$ to invert. Finally if we consider Fig. 22 we see we can remove only one mesh point at a time. Thus with this approach we can solve the three-dimensional problem by a sequence of scalar minimizations. It is a tedious but direct exercise to derive recurrence equations for all these approaches.

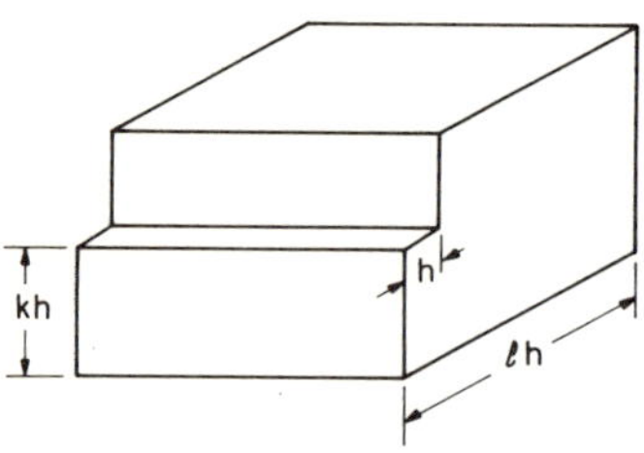

FIGURE 20

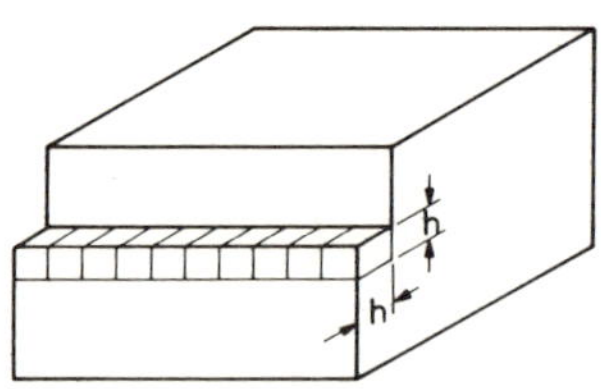

FIGURE 21

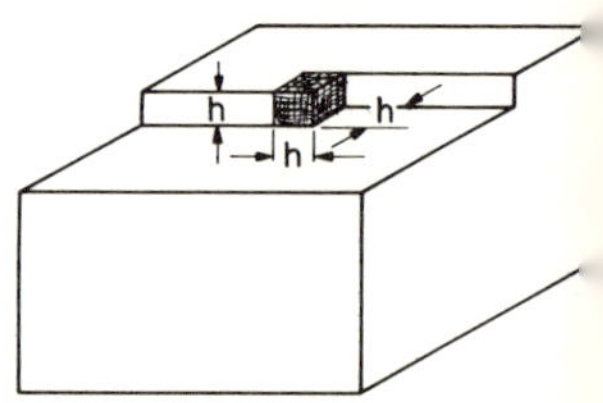

FIGURE 22

BIBLIOGRAPHY AND COMMENT

Section 2. For the potential and Poisson equations, various approaches have been suggested for the irregular region problem. See for instance

B. L. Buzbee, G. H. Golub, and C. W. Nielson, "On Direct Methods for Solving Poisson's Equations," *SJAM J. Num. Anal.*, Vol. 7, 1970, pp. 627–656.

Section 3. See

E. Angel, "Discrete Invariant Imbedding and Elliptic Boundary-Value Problems Over Irregular Regions," *J. Math. Anal. Appl.* Vol. 23, 1968, pp. 471–484.

Section 7. The Sturmian separation theorem is discussed in

R. Bellman, *Introduction to Matrix Analysis*, 2nd ed. McGraw-Hill, New York, 1970.

Section 8. See

S. G. Mikhlin and K. L. Smolitskly, *Approximate Methods for the Solution of Differential and Integral Equations*, Amer. Elsevier, New York, 1967.

Section 9. Further numerical results are contained in

E. Angel, "Dynamic Programming and Partial Differential Equations," Ph.D. Thesis, Univ. of Southern California, 1968.

Section 11. See the list of references in

R. Varga, *Matrix Iterative Analysis*, Prentice-Hall, Englewood Cliffs, New Jersey, 1962.

Section 12. See

E. Angel, *Invariant Imbedding and Three Dimensional Potential Problems*, Electronic Sciences Laboratory, Univ. of Southern California, USCEE-325, 1969.

Section 13. See

N. Distefano and J. Schujman, *Numerical Solution of Boundary-Value Problems in Structural Mechanics by Reduction to an Initial-Value Formulation*, Earthquake Engineering Research Center Rept. EERC 69-4, Univ. of California, Berkeley, 1969.

Section 14. The various relationship between Riccati equations and linearity are explored in

E. D. Denman, *Coupled Modes in Plasmas, Elastic Media and Parametric Amplifiers*, Amer. Elsevier, New York, 1970.

R. Redheffer, "Difference Equations and Functional Equations in *Modern Mathematics for the Engineer: Transmission-Line Theory*, E. F. Beckenbach, ed., 2nd Ser., McGraw-Hill, New York, 1962.

W. T. Reid, "Solution of A Riccati Matrix Differential Equation as Functions of Initial Values," *J. Math. Mech.*, Vol. 8, 1959, pp. 221–xxx.

W. T. Reid, *Riccati Differential Equations*, Academic Press, New York, (to appear).

A. McNabb and A. Schumitzky, "Factorization of Operators—III: Initial Value Methods for Linear Two-Point Boundary Value Problems," *J. Math. Anal. Appl.*, Vol. 31, 1970, pp. 391–406.

Section 15. See

E. Angel and R. Bellman, "Dynamic Programming and Reduction of Dimensionality for the Potential Equation," *J. Math. Anal. Appl.* (to appear).

Section 17. Another approach to the irregular region problem is to regard an irregular region as imbedded in a regular region. Constraints are used to fix the boundary conditions on the boundary of the irregular region which is now inside the chosen regular region. The advantage of this approach is that methods which work well for regular regions can be extended to very general problems. See

E. Angel, *Irregular Regions and Constrained Optimization*, Electronic Sciences Laboratory, Univ. of Southern California, USCEE 71–27, 1971.

B. L. Buzbee, F. W. Dorr, J. A. George and G. H. Golub, *The Direct Solution of the Discrete Poisson Equation on Irregular Regions*, Computer Sciences Department, Stanford Univ. STAN-CS-71-195, 1970.

Chapter 8

Special Computational Methods

1. Direct versus Iterative Methods

Our two basic approaches, dynamic programming and invariant imbedding, are both direct (i.e., noniterative) methods. As a consequence of these approaches we were required to invert sequences of symmetric matrices of general form. Hence, for a square region our methods require $O(N^4)$ multiplications and divisions. We argued that the relatively high number of operations was compensated for by a number of factors such as the ability to solve similar problems with almost no additional computation, the ease with which these methods can handle irregular regions, and the freedom from the choice of critical parameters.

On the other hand, the iterative finite difference techniques require only $O(N^2)$ operations per iteration and totally avoid matrix inversions. These are certainly important advantages in their turn. In this chapter we will derive a number of iterative methods directly from dynamic programming and invariant imbedding. Although we will show how these methods are equivalent to the standard methods, the approach we use will allow us to extend the iterative techniques to irregular region problems.

We will also discuss some modifications of our direct methods which will apply to constant coefficient equations. With these modifications we will be able to avoid matrix inversions.

Our example will again be Laplace's equation on the square. We use this example so that we can conveniently demonstrate the technique used to analyze iterative methods.

We will make extensive use of the characteristic values of the matrix Q, using Kronecker products to aid the analysis. Although these techniques are restricted to constant coefficient equations, we can justify their use on two accounts. First, it is through a thorough understanding of the constant coefficient equations that we will be able to approach more difficult problems. Second, these equations are important in their own right, as evidenced by the extensive body of literature on the potential, Poisson, and biharmonic equations.

2. The Characteristic Values of Q

Before we proceed further, we must dispense with a few mathematical preliminaries. We have seen that the analysis of all our methods depended on the positive definiteness of the matrix Q. At this point, then, it will be useful to examine the characteristic values of Q.

Let Q be of order N. We write Q as

$$Q = 2I - P, \tag{1}$$

thus defining the matrix P as

$$P = \begin{bmatrix} 0 & 1 & & \bigcirc \\ 1 & 0 & 1 & \\ & 1 & \ddots & 1 \\ \bigcirc & & 1 & 0 \end{bmatrix}. \tag{2}$$

Thus, if we find the characteristic values of P we will have the characteristic values of Q. Let the matrix F,

$$F = (f_{ij}), \tag{3}$$

be defined by

$$f_{ij} = \sin[ip\pi/(N+1)] \sin[jq\pi/(N+1)], \tag{4}$$

where p and q are arbitrary integers. Then, if

$$PF = (a_{ij}), \tag{5}$$

we have by direct multiplication

$$\begin{aligned} a_{ij} &= \{\sin[(i+1)\,p\pi/(N+1)] + \sin[(i-1)\,p\pi/(N+1)]\} \\ &\quad \times \sin[jq\pi/(N+1)] \\ &= 2\sin[i\pi p/(N+1)]\cos[p\pi/(N+1)]\sin[jq\pi/(N+1)]. \end{aligned} \tag{6}$$

Let D denote the diagonal matrix

$$D = \begin{bmatrix} v_1 & & & \\ & v_2 & & \\ & & \ddots & \\ & & & v_N \end{bmatrix}. \tag{7}$$

Then we find, setting

$$FD = (b_{ij}), \tag{8}$$

that

$$b_{ij} = \sin[i\pi p/(N+1)]\sin[j\pi q/(N+1)]\, v_j. \tag{9}$$

Thus,

$$PF = FD, \tag{10}$$

provided that

$$v_j = 2\cos[jp\pi/(N+1)]. \tag{11}$$

Since p is arbitrary, the characteristic values of P are given by

$$v_i = 2\cos[i\pi/(N+1)], \qquad i = 1, 2, \ldots, N. \tag{12}$$

and thus, the characteristic values, μ_j, of Q are the quantities

$$\mu_j = 2\{1 - \cos[j\pi/(N+1)]\}, \qquad j = 1, 2, \ldots, N. \tag{13}$$

3. Kronecker Product

In the previous chapter we briefly mentioned the Kronecker product. In this chapter we will use these products extensively. Let A be an N-dimensional matrix and B an M-dimensional matrix. The Kronecker product $A \otimes B$ is defined as the NM-dimensional matrix

$$C = A \otimes B = (a_{ij} B). \tag{1}$$

We can easily show that the Kronecker product thus defined has a number of convenient algebraic properties. For instance

$$A \otimes (B \otimes C) = (A \otimes B) \otimes C, \tag{2}$$

$$(A + B) \otimes (C + D) = A \otimes C + A \otimes D + B \otimes C + B \otimes D, \tag{3}$$

$$(A \otimes B)(C \otimes D) = (AC) \otimes (BD). \tag{4}$$

EXERCISE

Prove (2), (3), and (4).

4. Kronecker Sums

Let us consider the matrix C defined by

$$C = I \otimes A + B \otimes I. \tag{1}$$

If A has order N and B order M we will use the correct orders of the identity matrices in (1), so that C has order MN. We will assume that A and B are both real symmetric since this is the only case we will use later.

Let $\{\alpha_i\}$ and $\{\beta_j\}$ be the characteristic values of A and B respectively. We write A and B as

$$A = T\bar{A}T', \qquad B = S\bar{B}S', \tag{2}$$

where the prime denotes matrix transposition and

$$\bar{A} = \begin{bmatrix} \alpha_1 & & 0 \\ & \ddots & \\ 0 & & a_M \end{bmatrix}, \qquad \bar{B} = \begin{bmatrix} \beta_1 & & 0 \\ & \ddots & \\ 0 & & \beta_N \end{bmatrix}, \tag{3}$$

and

$$TT' = I, \qquad SS' = I. \tag{4}$$

We now use (2) in (1) and apply the properties of the Kronecker product noted in Section 3,

$$\begin{aligned} C &= I \otimes (T\bar{A}T') + S\bar{B}S') \otimes I \\ &= (SS') \otimes (T\bar{A}T') + (S\bar{B}S') \otimes (TT') \\ &= (S \otimes T)(I \otimes \bar{A})(S' \otimes T') + (S \otimes T)(\bar{B} \otimes I)(S' \otimes T') \\ &= (S \otimes T)(I \otimes \bar{A} + \bar{B} \otimes I)(S \otimes T)'. \end{aligned} \tag{5}$$

We can directly verify that

$$(S \otimes T)(S \otimes T)' = I \tag{6}$$

and $I \otimes \bar{A} + \bar{B} \otimes I$ is diagonal. Thus, the characteristic values of C are the diagonal entries of $I \otimes A + B \otimes I$, the MN numbers $\alpha_i + \beta_j$. We will use this result extensively.

EXERCISE

Show that the block tridiagonal matrix

$$D = \begin{bmatrix} A & B & & & \\ C & A & B & & \\ & \ddots & \ddots & \ddots & \\ & & C & A & B \end{bmatrix}$$

can be written in the form

$$D = [I \otimes A + L \otimes C + U \otimes B],$$

where L and U are defined as

$$L = \begin{bmatrix} 0 & & & & \bigcirc \\ 1 & 0 & & & \\ & 1 & \ddots & & \\ & & \ddots & & \\ \bigcirc & & & 1 & 0 \end{bmatrix}, \qquad U = \begin{bmatrix} 0 & 1 & & & \bigcirc \\ & 0 & 1 & & \\ & & \ddots & \ddots & \\ & & & & 1 \\ \bigcirc & & & & 0 \end{bmatrix}.$$

5. An Example

Consider our discretized version of the potential equation

$$u_{i+1} - [4I - P]u_i + u_{i-1} - r_i = 0, \tag{1}$$

where we have replaced Q by $2I - P$ as in Section 2. We will assume that the region is a square. Thus u_i is of order N, u_0 and u_{N+1} are known. We saw in Chapter 6 that (1) could be written as

$$\begin{bmatrix} 4I-P & -I & & \bigcirc \\ -I & 4I-P & & \\ & & \ddots & \\ & & & -I \\ \bigcirc & & -I & 4I-P \end{bmatrix} \begin{bmatrix} u_1 \\ u_2 \\ \vdots \\ \\ u_N \end{bmatrix} = \begin{bmatrix} u_0 + r_1 \\ r_2 \\ \vdots \\ u_{N+1} \\ u_{N+1} - r_N \end{bmatrix},$$

a relation of the form

$$Au = r, \tag{3}$$

where these quantities are of order N^2.

Closer investigation shows that the matrix A has the structure

$$A = -I \otimes P - P \otimes I + 4I \otimes I. \tag{4}$$

Thus we can write the solution to (1) as

$$u = [-I \otimes P - P \otimes I + 4I \otimes I]^{-1} r. \tag{5}$$

If we write P as

$$P = FDF, \tag{6}$$

where

$$D = \begin{bmatrix} \nu_1 & & & \\ & \nu_2 & & \\ & & \ddots & \\ & & & \nu_N \end{bmatrix} \tag{7}$$

then we see that (5) becomes

$$u = [F \otimes F][-I \otimes D - D \otimes I + 4I \otimes I]^{-1} [F' \otimes F'] r. \tag{8}$$

However the matrix $[-I \otimes D - D \otimes I + 4I \otimes I]^{-1}$ is a diagonal matrix whose elements are the numbers $1/(4 - \nu_i - \nu_j)$. Since we already have shown that

$$\mu_i = 2\cos[i\pi/(N+1)] \tag{9}$$

and

$$F = \{\sin[i\pi/(N+1)] \sin[j\pi/(N+1)]\}, \tag{10}$$

(8) gives a direct method of solving the potential equation on rectangular regions. In the literature, this technique is known as the "tensor product" method.

EXERCISE

Show that the tensor product method requires $O(N^3)$ operations.

6. Another Direct Method

Suppose we are solving either the potential equation or the Poisson equation on a rectangular region. We saw in Chapter 6 that the invariant imbedding approach led to the following equations for the vector u_N

$$\begin{aligned} R_{i+1} &= [2I + Q - R_i]^{-1}, \qquad i = 0, 1, \ldots, \quad R_0 = 0, \\ s_{i+1} &= R_{i+1}(s_i + r_i), \qquad i = 0, 1, \ldots, \quad s_0 = c, \end{aligned} \tag{1}$$

and

$$\begin{aligned} U_{i+1} &= R_{i+1} U_i, \\ & i = n, n+1, \ldots, N, \\ p_{i+1} &= p_i + U_i s_i, \\ p_n &= 0, \qquad U_n = R_n. \end{aligned} \tag{2}$$

Finally we obtained the desired solution in the form

$$u_n = U_N d + p_N. \tag{3}$$

If we write Q as

$$Q = FDF, \tag{4}$$

where D is diagonal and F is the symmetric orthogonal matrix we introduced in Section 2, then we see inductively that

$$R_i - F\bar{R}_i F \tag{5}$$

and

$$U_i = F\bar{U}_i F, \tag{6}$$

where $\bar{R}_i$ and $\bar{U}_i$ are diagonal. Let us define vectors $\bar{s}_i$ and $\bar{p}_i$ by

$$\bar{s}_i = Fs_i, \qquad \bar{p}_i = Fp_i. \tag{7}$$

Now it is easy to verify that these new quantities satisfy the equations

$$\begin{aligned} \bar{R}_{i+1} &= [D - \bar{R}_i]^{-1}, \qquad \bar{R}_1 = 0, \quad i = 0, 1, \ldots, \\ \bar{s}_{i+1} &= \bar{R}_i(\bar{s}_i + F_i r_i), \qquad \bar{s}_1 = Fc, \quad i = 0, 1, \ldots, \end{aligned} \tag{8}$$

and

$$\begin{aligned} \bar{U}_{i+1} &= \bar{R}_i \bar{U}_i, \\ \bar{p}_{i+1} &= \bar{p}_i + \bar{U}_i \bar{s}_i, \qquad i = n, n+1, \ldots, \\ \bar{U}_n &= \bar{R}_n, \qquad p_n = 0. \end{aligned} \tag{9}$$

Finally we obtained the desired solution by

$$u_n = F\bar{U}_N Fd + F\bar{p}_N. \tag{10}$$

Since all the matrices except F are diagonal, we have eliminated all the matrix inversions, and since D and F are known we can easily carry out these computations.

EXERCISE

Show that this method requires $O(N^2)$ operations for the potential equation and $O(N^3)$ for Poisson's equation.

7. Diagonal Decomposition

In Chapter 5 we discretized the variational problem corresponding to the potential equation in the following fashion

$$J(u) = \sum_{R=1}^{N} [(Qu_R, u_R) - (2r_R, u_R) + s_R + (u_R - u_{R-1}, u_R - u_{R-1})], \tag{1}$$

and we found the quantities u_R by applying dynamic programming to (1). Using the principle of optimality we obtained the functional equation for the truncated minimization problem,

$$f_R(v) = \min_{u_R} [(Qu_R, u_R) - (2r_R, u_R) + s_R + (u_R - v, u_R - v) + f_{R+1}(u_R)], \tag{2}$$

with

$$v = u_{R-1}. \tag{3}$$

The function $f_1(u_0)$ provides the minimum of (1). It is clear by observation by (2) that the matrix Q provides the only interaction between the components of the state vector, u_R. If we were to go through the development of Chapter 5 again, we would see that it is precisely this interaction which forces us to carry out a sequence of matrix inversions.

Suppose we have an approximation to the solution, that is, a set of vectors $\{u_r^{(0)}\}$. We can write Q as

$$Q = 2I - P, \tag{4}$$

where P is again the matrix

$$P = (p_{ij}), \qquad p_{ij} = \begin{cases} 1, & |i-j| = 1, \\ 0, & \text{otherwise.} \end{cases} \tag{5}$$

Thus, by (4)

$$(Qu_R, u_R) = (2u_R, u_R) - (Pu_R, u_R). \tag{6}$$

If $u_R^{(0)}$ is close to u_R,

$$(P(u_R - u_R^{(0)}), u_R - u_R^{(0)}) \cong 0, \tag{7}$$

whence

$$(Pu_R, u_R) \cong (2Pu_R^{(0)}, u_R) - (Pu_R^{(0)}, u_R^{(0)}). \tag{8}$$

Note that our approximation has the property that

$$\frac{d}{du_R}(Pu_R, u_R) \cong \frac{d}{du_R}[(2Pu_R^{(0)}, u_R) - (Pu_R^{(0)}, u_R^{(0)})], \tag{9}$$

an important property which we would not have if we used the simpler approximation

$$(Pu_R, u_R) \cong (Pu_R^{(0)}, u_R). \tag{10}$$

We now use (8) in (2), and we get the approximate formula

$$\begin{aligned} f_R(v) = \min_{u_R} [(2u_R, u_R) - (2Pu_R^{(0)}, u_R) + (Pu_R^{(0)}, u_R^{(0)}) - (2r_R, u_R) + s_R \\ + (u_R - v, u_R - v) + f_{R+1}(u_R)], \end{aligned} \tag{11}$$

or

$$f_R(v) = \min_{u_R} [(2u_R, u_R) - (2\tilde{r}_R, u_R) + \tilde{s}_R + (u_R - v, u_R - v) + f_{R+1}(u_R)], \tag{12}$$

where

$$\tilde{r}_R = r_R + Pu_R^{(0)}, \qquad \tilde{s}_R = s_R + (Pu_R^{(0)}, u_R^{(0)}). \tag{13}$$

The function $f_R(v)$ defined by (12) is clearly quadratic in v. Thus if we set

$$f_R(v) = (A_R v, v) - (2b_R, v) + c_R, \tag{14}$$

we find, by the same procedure we used in Chapter 5, that the minimizing u_R is given by

$$u_R = [3I + A_{R+1}]^{-1}(v + \tilde{r}_R + b_{R+1}), \tag{15}$$

and that the functions A_R and b_R satisfy

$$\begin{aligned} A_R &= I - [3I + A_{R+1}]^{-1}, \\ b_R &= [I - A_R]\,(b_{R+1} + \tilde{r}_R), \end{aligned} \tag{16}$$

with

$$A_N = I, \qquad b_N = u_{N+1}. \tag{17}$$

It is clear inductively that A_R as defined by (16) is diagonal. Thus no matrix inversions are necessary. In fact to get A_R from A_{R+1} only N scalar divisions are necessary and the solution of (15)–(17) requires only $2N^2$ multiplications and divisions. The vectors $\{u_R\}$ given by the above procedure will of course only be a next approximation and we must iterate on (15) and (16). Using (13) we can rewrite our equations in the form

$$A_R = I - [3I + A_{R+1}]^{-1}, \qquad A_N = I, \tag{18}$$

$$b_R^{(k+1)} = [I - A_R]\,(b_{R+1}^{(k+1)} + r_R + Pu_R^{(k)}), \qquad b_N^{(k+1)} = u_{N+1}, \tag{19}$$

and

$$u_{R+1}^{(k+1)} = [I - A_R]\,u_R^{(k+1)} + b_R^{(k+1)}. \tag{20}$$

Since the equation for A_R is independent of the boundary conditions and the approximation, we need only solve it once.

We can show that the method defined by (18)–(20) converges. However we will delay the proof until we have developed a number of iterative methods from the invariant imbedding approach. It will be easier to prove convergence from this approach.

8. Point Iterative Methods

The discretization of the potential equation led us to the matrix vector difference equation

$$u_{R+1} - 2u_R + u_{R-1} - Qu_R + r_R = 0, \tag{1}$$

where u_0 and u_N are known. Suppose we now use the fact that Q can be written as

$$Q = 2I - P. \tag{2}$$

Now we can write (1) as

$$u_R = \tfrac{1}{4}[Pu_R + u_{R+1} + u_{R-1} + r_R], \tag{3}$$

which immediately suggests the simple iterative scheme

$$u_R^{(k+1)} = \tfrac{1}{4}[Pu_R^{(k)} + u_{R+1}^{(k)} + u_{R-1}^{(k)} + r_R], \tag{4}$$

where $\{u_R^{(0)}\}$ are initial guesses. Note that although (4) retains our matrix vector notation, each component of $u_R^{(k+1)}$ is computed independently from each other component. Thus, all our computations are scalar computations. The procedure in (4) is known as the "point Jacobi method."

If (4) converges, it will converge to the solution of (3). Let the error, $e_R^{(k)}$ be defined by

$$e_R^{(k)} = u_R - u_R^{(k)}. \tag{5}$$

We immediately find that $e_R^{(k)}$ satisfies the difference equation

$$e_R^{(k+1)} = \tfrac{1}{4}[Pe_R^{(k)} + e_{R+1}^{(k)} + e_{R-1}^{(k)}], \tag{6}$$

with

$$e_0^{(k)} = e_{N+1}^{(k)} = 0, \tag{7}$$

for all k. Let us define the N^2 dimensional vector $e^{(k)}$ as

$$e^{(k)} = [e_i^{(k)}]. \tag{8}$$

Using Kronecker products, we can now write (6) as

$$e^{(k+1)} = \tfrac{1}{4}[P \otimes I + I \otimes P]\,e^{(k)}. \tag{9}$$

If we let T denote the matrix

$$T = \tfrac{1}{4}[P \otimes I + I \otimes P], \tag{10}$$

it is clear that

$$e^{(k)} = T^k e^{(0)}, \tag{11}$$

so that

$$\lim_{k\to\infty} e^{(k)} \to 0, \tag{12}$$

if and only if

$$\rho(T) < 1. \tag{13}$$

Fortunately, we have shown that the characteristic values of P are the numbers

$$\nu_i = 2\cos[i\pi/(N+1)]. \tag{14}$$

Hence we find that

$$\rho(T) = \cos[\pi/(N+1)]. \tag{15}$$

Thus, the method will always converge. However, for N large

$$\cos[\pi/(N+1)] \cong 1 - \tfrac{1}{2}[\pi/(N+1)]^2. \tag{16}$$

This means that for fine discretizations the convergence will be extremely slow.

The point Jacobi method requires that we wait until the end of the kth iteration before we use the new values we have just computed. If we assume that we always iterate in the order of increasing x and y and that we use new estimates as soon as possible, we get the Gauss–Seidel method,

$$u_R^{(k+1)} = \tfrac{1}{4}[Lu_R^{(k+1)} + Uu_R^{(k)} + u_{R-1}^{(k+1)} + u_{R+1}^{(k)} + r_R], \tag{17}$$

where L and U are defined by

$$L = \begin{bmatrix} 0 & & & & \bigcirc \\ 1 & 0 & & & \\ & 1 & \ddots & & \\ & & \ddots & & \\ \bigcirc & & & 1 & 0 \end{bmatrix}, \qquad U = \begin{bmatrix} 0 & 1 & & & \bigcirc \\ & 0 & 1 & & \\ & & \ddots & \ddots & \\ & & & & 1 \\ \bigcirc & & & & 0 \end{bmatrix}. \tag{18}$$

Equivalently, L and U are a strictly lower triangular matrix and strictly upper triangular matrix such that

$$L + U = P. \tag{19}$$

EXERCISES

1. Show that the significant matrix T for the Gauss–Seidel method is given by

$$T = [I \otimes (4I - L) - L \otimes I]^{-1}[I \otimes U + U \otimes I].$$

2. Show that $\rho(T) = \{\cos[\pi/(N+1)]\}^2$.

9. The Successive Overrelaxation Method

We saw that for both the Jacobi and Gauss–Seidel methods, the moduli of the significant matrices will be close to one if the grid spacing was small. To get around this problem, we can use an acceleration parameter.

Let $\bar{u}_R^{(k+1)}$ denote the value yielded by a Gauss–Seidel iteration,

$$\bar{u}_R^{(k+1)} = \tfrac{1}{4}[Lu_R^{(k+1)} + Uu_R^{(k)} + u_{R-1}^{(k+1)} + u_{R+1}^{(k)} + r_R]. \tag{1}$$

Then we will obtain the new iterate, $u_R^{(k+1)}$, by taking a weighted average of this value and the last value,

$$u_R^{(k+1)} = \omega\bar{u}_R^{(k+1)} + (1-\omega)\,u_R^{(k)}, \tag{2}$$

where ω is our acceleration parameter. Eliminating $\bar{u}_R^{(k+1)}$ from (1) and (2) we find that the point relaxation method is defined by

$$[4I - \omega L]\,u_R^{(k+1)} - \omega u_{R-1}^{(k+1)} = [4(I-\omega)I + \omega U]\,u_R^{(k)} + \omega u_{R-1}^{(k)} + \omega r_R. \tag{3}$$

The reader can easily verify that the significant matrix is

$$T = [I \otimes (4I - \omega L) - \omega L \otimes I]^{-1}[I \otimes (4(1-\omega)I + \omega U) + \omega U \otimes I]. \tag{4}$$

The characteristic values of T are the solutions of the determinantal equation

$$|T - \lambda I| = 0. \tag{5}$$

In the exercises we show how this equation may be solved. The result is that the ith characteristic value of T, λ_i, must satisfy the relation

$$\lambda_i + \omega - 1 = \mu_i \omega \lambda_i^{1/2}/2, \tag{6}$$

where again μ_i is the ith characteristic value of P. Note that for $\omega = 1$, we find the characteristic values of the Gauss–Seidel method.

We must now find for what value of ω the modulus of T is minimal, that is we must find

$$\rho(T) = \min_{\omega} \max_{i} |\lambda_i|. \tag{7}$$

Considering the complex mapping defined by (6), Young has shown that the optimal ω is determined by

$$\omega^2 \nu^2 = 4(\omega - 1), \tag{8}$$

where ν is the largest characteristic value of P. The method thus defined has been called "successive overrelaxation" since the optimal ω satisfies the relation

$$1 \leqslant \omega < 2. \tag{9}$$

Finally we find that $\rho(T)$ is given by

$$\rho(T) = \frac{1-[1-(\rho/2)^2]^{1/2}}{1+[1-(\rho/2)^2]^{1/2}}, \tag{10}$$

where v is the largest characteristic value of P. For N large we have

$$\rho(T) \cong 1 - 2\sqrt{2}[\pi/(N+1)]. \tag{11}$$

If we define a rate of convergence by

$$r(T) = -\log \rho(T), \tag{12}$$

we see that for N large that the rate of convergence of the Gauss–Seidel method is twice that of the Jacobi method while the rate of convergence of the successive overrelaxation method is twice the square root of the Gauss–Seidel method. Thus, for N large, the successive overrelaxation is vastly superior to the other iterative methods we have just discussed.

EXERCISES

1. Show that (1) and (2) define a point iterative method. That is, all the required computations are scalar computations.

2. Let the matrices R and S be defined by

$$R = \begin{bmatrix} I & & & \bigcirc \\ & \lambda^{1/2} I & & \\ & & \ddots & \\ \bigcirc & & & \lambda^{(N-1)/2} I \end{bmatrix}, \quad S = \begin{bmatrix} 1 & & & \bigcirc \\ & \lambda^{1/2} & & \\ & & \ddots & \\ \bigcirc & & & \lambda^{(N-1)/2} \end{bmatrix},$$

so that P is of order N^2 and S is of order N. Show that if T is as defined by (4), then premultiplication of T by $[I \otimes S]P$ and postmultiplication by $[[I \otimes S]P]^{-1}$ yields the characteristic values of T. (See the article by D. Young listed in the bibliography at the end of the chapter.)

10. Block Iterative Methods

The preceeding iterative methods all had the property that to compute some component of $u_R^{(k+1)}$, we had only to do arithmetic operations involving the present iterates at adjoining points. The idea of the block iterative methods is to iterate on a number of points simultaneously. We will consider only the row iterative methods. The column iterative methods will be left to the exercises.

We start with our original system and again decompose Q so that we have

$$-u_{R+1} + (4I - P)u_R - u_{R-1} + r_R = 0. \tag{1}$$

We can use this equation iteratively in the form

$$-u_{R+1}^{(k+1)} + 4u_R^{(k+1)} - u_{R-1}^{(k+1)} = Pu_R^{(k)} + r_R. \tag{2}$$

This new problem is still a two-point boundary value problem with u_0 and u_{N+1} known. We seek a solution of the form

$$u_{R+1}^{(k+1)} = A_R u_R^{(k+1)} + b_R^{(k+1)}, \tag{3}$$

and we find by the usual technique the recurrence equations

$$\begin{aligned} A_{R-1} &= [4I - A_R]^{-1}, \\ b_{R-1}^{(k+1)} &= A_{R-1}(b_R^{(k+1)} + r_R + Pu_R^{(k)}), \end{aligned} \tag{4}$$

with

$$A_N = 0, \qquad b_N^{(k+1)} = u_{N+1}. \tag{5}$$

It is clear inductively that the matrices A_i are diagonal, independent of the iteration number k, and that all the diagonal elements of A_i are the same. Thus, we can determine all the matrices A_R with N divisions. Each iteration then amounts to finding new $b_i^{(k)}$ and $u_i^{(k)}$, a procedure which requires $2N^2$ multiplications.

The above procedure amounts to a simultaneous iteration of the points of u_R in the Jacobi manner. That is we assume that values in adjoining rows are given at the kth iteration while those in the given row are evaluated implicitly at the $k+1$st iteration. It is also easy to show that this procedure is equivalent to diagonal decomposition.

We can investigate the convergence of this method through the same techniques we employed previously. The error satisfies

$$-e_{R+1}^{(k+1)} + 4e_R^{(k+1)} - e_{R-1}^{(k+1)} = Pe^{(k)}. \tag{6}$$

It follows immediately that the significant matrix is

$$T = [4I \otimes I - I \otimes P]^{-1}[P \otimes I] \tag{7}$$

and thus

$$\rho(T) = (4 - \nu)/\nu, \tag{8}$$

where again ν is the largest characteristic value of P. For N, large, we have

$$\rho(T) \cong 1 - 2[\pi^2/(n + 1)]. \tag{9}$$

Thus, the block Jacobi method has a rate of convergence twice that of the point Jacobi method. Unfortunately the block process requires twice as many operations per iteration so that we have not really gained a great deal over the point process. Similar results hold for the block Gauss–Seidel and block overrelaxation method; neither has large advantage over the corresponding point method.

EXERCISES

1. Derive the block Gauss–Seidel and block overrelaxation methods. Show that the significant matrix for the overrelaxation procedure is given by

$$T = [(4I - \omega L) \otimes I - I \otimes P]^{-1} [(4(1 - \omega) I + \omega U)] \otimes I + (1 - \omega) I \otimes P].$$

2. Show that the "column Jacobi method" is described by the equation

$$[2I + Q] u_R^{(k+1)} = u_{R+1}^{(k)} + u_{R-1}^{(k)} + r_R.$$

Analyze this method.

3. Show that the row Jacobi method is equivalent to diagonal decomposition.

11. Alternating-Direction Implicit Methods

The alternating-direction implicit method of Peaceman and Rachford is similar to a column iteration followed by a row iteration. Since this method arises from the study of parabolic equations, we will be content at the moment to describe the method. In our notation, the Peaceman–Rachford scheme becomes

$$[I + cQ] u_R^{(k+1/2)} = c[u_{R+1}^{(k)} + u_{R-1}^{(k)} + r_R] + (1 - 2c) u_R^{(k)}, \tag{1}$$

and

$$c u_{R+1}^{(k+1)} - (2c + 1) u_R^{(k+1)} + c u_{R-1}^{(k+1)} = [I - cQ] u_R^{(k+1/2)} + c r_R, \tag{2}$$

where c is a scalar parameter which may vary with k. Since (1) is a tridiagonal system we can solve it in exactly the same way we have solved the potential equation, i.e., see Chapter 3, Section 12. All the equations will be scalar equations. We have already seen that the solution of (2) also amounts to simple scalar computations. It is easy to show that the solution of (1) and (2) require $O(4N^2)$ operations.

If again we let $e^{(k)}$ denote the N^2 dimensional vector

$$e^{(k)} = [e_R^{(k)}], \tag{3}$$

we have

$$e^{(k+\frac{1}{2})} = [(I + cQ) \otimes I]^{-1}[(1-2c)I \otimes I + cP \otimes I]e^{(k)}, \tag{4}$$

and

$$e^{(k+1)} = [cP \otimes I - (2c+1)I \otimes I]^{-1}[I \otimes (I - cQ)]e^{(k+\frac{1}{2})}. \tag{5}$$

Combining these results we find the significant matrix to be given by

$$\begin{aligned} T = {} & [cP \otimes I - (2c+1)I \otimes I]^{-1}[I \otimes (I - cQ)] \\ & \times [(I - cQ) \otimes I]^{-1}[(1-2c)I \otimes I + cP \otimes I]. \end{aligned} \tag{6}$$

After some manipulation we find that $\rho(T)$ is given by

$$\rho(T) = \left[\frac{1-(\mu-2)c}{1+(\mu-2)c}\right]^2, \tag{7}$$

where μ is a characteristic value of P. If we now let c vary with k, the problem becomes one of finding a sequence $\{c_k\}$ to minimize the absolute value of

$$f(M,K) = \prod_{k=1}^{K} \frac{1-(\mu-2)c_k}{1-(\mu-2)c_k}. \tag{8}$$

Although the $\{c_k\}$ can be determined for rectangles, the general problem is very difficult.

12. Discussion

For a problem such as the potential equation over a rectangle, it is clear that a number of direct and iterative methods can be used very effectively. Unfortunately as soon as the region gets irregular, or the equation is allowed to have variable coefficients, these methods become either inapplicable or difficult to use. For instance, it is well known that the alternating direction implicit methods do not work over some very simple irregular regions. We also saw that while a sophisticated iterative method was necessary to have rapid convergence, the determination of optimal parameters was very difficult. For these reasons we will not discuss these methods any further. The reader should realize, however, that these methods are applicable to problems we discuss later in the book.

BIBLIOGRAPHY AND COMMENT

Section 1. For extensive lists of references see

R. Varga, *Matrix Iterative Analysis*, Prentice-Hall, Englewood Cliffs, New Jersey, 1962.
F. W. Dorr, "The Direct Solution of the Discrete Poisson Equation on a Rectangle, *SIAM Rev.*, Vol. 12, 1970, pp. 248–263.
G. E. Forsythe and W. R. Wasow, *Finite Difference Methods for Partial Differential Equations*, Wiley, New York, 1960.
B. Carnahan, H. A. Luther, and J. O. Wilkes, *Applied Numerical Methods*, Wiley, New York, 1969.

Section 2. The proof follows that of D. Young. See

J. Todd, ed., *Survey of Numerical Analysis*, McGraw-Hill, New York, 1962.

Section 3, Section 4. See

R. Bellman, *Introduction to Matrix Analysis*, 2nd ed., McGraw-Hill, New York, 1970.

Section 5. See

R. E. Lynch, J. R. Rice, and D. H. Thomas, "Tensor Product Analysis of Partial Difference Equations," *Bull. Amer. Math. Soc.*, Vol. 70, 1964, pp. 378–384.
R. E. Lynch, J. R. Rice, and D. H. Thomas, "Direct Solution of Partial Difference Equations by Tensor Product Methods," *Numer. Math.*, Vol. 6, 1964, pp. 185–199.

A useful alternative to the tensor product method is provided by a condensation theorem of J. Williamson (*Bull. Amer. Math. Soc.*, Vol. 37, 1931, pp. 585–590): If $A = [A_{ij}]$ is an $M \times M$ block matrix with each A_{ij} an $N \times N$ matrix and if the A_{ij} commute (in particular, if they are all rational functions r_{ij} of a matrix A, with characteristic values $\alpha_1, \alpha_2, \cdots, \alpha_N$), then the characteristic values of A are the N sets of characteristic values of the $M \times M$ matrices $[r_{ij}(\alpha_k)]$, $k = 1, 2, \cdots, N$. For details see, for example,

J. Todd, "Numerical Analysis," Chap. 7 in *Handbook of Physics*, E. U. Condon and H. Odishaw, eds., 2nd ed., Part 1, 1967.

Section 6. There are a number of other methods which depend on knowledge of the characteristic values of Q. See

R. W. Hockney, "A Fast Direct Solution of Poisson's Equation using Fourier Analysis," *J. Assoc. Comput. Mach.*, Vol. 12, 1965, pp. 95–113.
B. L. Buzbee, G. H. Golub, and C. W. Nielson, *The Method of Odd-even Reduction and Factorization with Applications to Poisson's Equation*, Los Alamos Scientific Laboratory Rept. LA-4141, 1969.
B. L. Buzbee, G. H. Golub, and C. W. Nielson, "On Direct Methods for Solving Poisson's Equation," *SIAM. J. Numer. Anal.* Vol. 7, 1970, pp. 627–656.

Section 7. See

D. C. Collins and E. Angel, "The Diagonal Decomposition Technique Applied to the Dynamic Programming Solution of Elliptic Partial Differential Equations," *J. Math. Anal. Appl.*, Vol. 33, 1971, pp. 467–481.

D. C. Collins and A. Lew, "Dimensional Approximation in Dynamic Programming by Structural Decomposition," *J. Math. Anal. Appl.*, Vol. 30, 1970, pp. 375–384.

R. E. Larson, *State Increment Dynamic Programming*, Amer. Elsevier, New York, 1968.

Section 8. See

E. Angel, "Invariant Imbedding, Difference Equations and Elliptic Boundary Value Problems," *J. Comput. System Sci.*, Vol. 4, 1970, pp. 473–491.

Section 9. See

D. Young, "Iterative Methods for Solving Partial Difference Equations of Elliptic Type," *Trans. Amer. Math. Soc.*, Vol. 76, 1954, pp. 92–111.

Section 10. See any of the references for Section 1 and Section 2. Also

E. H. Cuthill and R. S. Varga, "A Method of Normalized Block Iteration," *J. Assoc. Comput. Mach.*, Vol. 6, 1959, pp. 236–244.

Section 11. See

D. W. Peaceman and H. H. Rachford Jr., "The Numerical Solution of Parabolic and Elliptic Differential Equations," *J. Soc. Indust. Appl. Math.*, Vol. 3, 1955, pp. 28–41.

J. Douglas, Jr., "On the Numerical Integration of $\partial^2 u/\partial x^2 + \partial^2 u/\partial y^2 = \partial u/\partial t$ by Implicit Methods," *J. Soc. Indust. Appl. Math.*, Vol. 3, 1955, pp. 42–65.

G. Birkhoff and R. S. Varga, "Implicit Alternating Direction Methods," *Trans. Amer. Math. Soc.*, Vol. 92, 1959, pp. 13–24.

Chapter 9

Unconventional Difference Methods

1. Introduction

In previous chapters we presented some approaches to the computational solution of partial differential equations based upon the use of dynamic programming and invariant imbedding. The algorithms obtained in this way were quite distinct from those obtained by a conventional use of finite difference methods. In this chapter we wish to discuss briefly some unconventional finite difference methods which are suggested by dynamic programming and invariant imbedding.

2. Invariant Imbedding

A characteristic partial differential equation in the theory of invariant imbedding is

$$r_t = g(r)r_x + h(r), \qquad r(x,0) = k(x). \tag{1}$$

The function $r(x, t)$ represents a reflected flux. Stratification plus "particle counting" yields in many physical processes an approximate relation for this reflected flux

$$r(x, t + \Delta) = r(x + g(r)\Delta, t) + h(r)\Delta + O(\Delta^2) \tag{2}$$

from which (1) follows upon letting Δ tend to zero.

It is of some interest then to reverse the process and examine the possibility of using (2) to obtain numerical values of $r(x, t)$. This is a finite difference relation of unconventional form.

3. The Equation $u_t = uu_x$

To illustrate the ideas, let us consider the equation

$$u_t = uu_x, \qquad u(x, 0) = g(x), \qquad -\infty < x < \infty. \tag{1}$$

This is an excellent example against which to test various proposed computational schemes since there exists a simple implicit solution for u,

$$u = g(x + ut). \tag{2}$$

From (2) we have

$$u_t = g'[u + tu_t] = g'u/(1 - tg'), \qquad u_x = g'[1 + tu_x] = g'/(1 - tg'), \tag{3}$$

whence (1) follows.

The solution exists as long as $1 - tg' \neq 0$. Hence we can expect that for some x there will be a first t value such that $1 - tg' = 0$. At this point there will be a "shock." In transport processes this corresponds to a "critical length."

4. Approximating Finite Difference Equation

Pursuing the idea of Section 2, let us consider the relation

$$v(x, t + \Delta) = v(x + v(x, t)\Delta, t), \qquad v(x, 0) = g(x), \tag{1}$$

where $-\infty < x < \infty$ and t assumes the values $0, \Delta, 2\Delta, \ldots$. For the purposes of computation in this test case let us take $g(x)$ to be periodic of period one, whence $v(x, t)$ will be periodic of period one for each t.

There are several approaches that can be followed to use (1) to calculate $v(x, t)$ recurrently starting from $g(x) = v(x, 0)$. The first is to store $v(x, t)$ for each t at a set of grid points $x = k\delta$, $k = 0, 1, \ldots, M$, where $M\delta = 1$. The right-hand side of (1) is then calculated at points where $x + v(x, t)\,\Delta$

is not a grid point by means of an interpolation scheme, e.g., linear interpolation.

Alternatively, we can store the function $u(x, t)$ for each t by means of some analytic representation, a trigonometric polynomial, an ordinary polynomial, a spline function; by means of differential approximation; and so forth.

Some sample calculations will be found in the references cited at the end of the chapter.

EXERCISES

1. Show that (1) is a finite difference approximation which preserves positivity and boundedness.

2. Obtain analogous relations for the partial differential system

$$u_t = uu_x + vu_y, \qquad v_t = uv_x + vv_y.$$

5. Convergence

The question arises as to whether the limit of $v(x, t) \equiv v(x, t, \Delta)$ as $\Delta \to 0$ exists and is the solution of (3.1). There are several different problems here. The first, which consists of two parts, involves starting with (4.1) and showing first that $v(x, t, \Delta)$ has a limit as $\Delta \to 0$,

$$\lim_{\Delta \to 0} v(x, t, \Delta) = u(x, t), \tag{1}$$

and secondly that this limit function satisfies (3.1). The second involves starting with the existence and uniqueness of a well-behaved solution of (3.1) and demonstrating that an approximation to this solution is furnished by the solution of (4.1).

It is interesting to note that this second problem, of principal interest to us here, is a problem in stability theory, the stability of the solution of (4.1).

Let $u(x, t)$ be the solution of (3.1), where $g(x)$ is taken to be sufficiently differentiable. Then

$$\begin{aligned} u(x, t + \Delta) &= u(x, t) + \Delta u_t + (\Delta^2/2)\, u_{tt} + \cdots, \\ u(x + u\Delta, t) &= u(x, t) + \Delta u u_x + (\Delta^2/2)\, u^2 u_{xx} + \cdots. \end{aligned} \tag{2}$$

Hence

$$u(x, t + \Delta) = u(x + u\Delta, t) + k(x, t)\, \Delta^2, \tag{3}$$

where $|k(x, t)| \leqslant k_1$ for $0 \leqslant x \leqslant 1$, $0 \leqslant t \leqslant T$.

Comparing this relation with (4.1), we see that $v \cong u$, provided that (4.1) is a stable equation for $\Delta \ll 1$. This can be demonstrated for a suitable (x, t) region.

6. Improvement of Accuracy

There are several ways in which we can improve the accuracy of approximation. One immediate way is to decrease the size of Δ. This has the disadvantage of increasing computing time. This may be mitigated to some extent by combining with a deferred passage to the limit approach.

Alternately, we can use a difference approximation of higher order. Consider, for example, the relation

$$v(x, t+\Delta) = v\big(x + v(x + v(x,t)\Delta, t)\Delta, t\big), \tag{1}$$

where $t = 0, \Delta, \ldots$, with $v(x, 0) = g(x)$.

Use of the original partial differential equation shows that this algorithm is accurate to $O(\Delta^3)$. In practice it appears that it is preferable to reduce Δ rather than use a more complicated difference relation, such as that in (1).

EXERCISES

1. Consider the equation $u_t = u_{xx}$. Show that $u(x, t+\Delta) = [u(x+\delta, t) + u(x-\delta, t]/2$ yields a consistent difference approximation provided that Δ and δ are suitably related. Show that the relation preserves positivity and boundedness.

2. Consider the recurrence relation

$$u(x, t+\Delta) = \sum_{i=1}^{N} w_i[u(x + a_i\delta, t) + u(x - a_i\delta, t)]\,.$$

Can one choose the parameters N, w_i, a_i and δ so that the relation is consistent with $u_t = u_{xx}$ to order Δ^k for any $k \geqslant 1$, with the further proviso that $w_i, a_i \geqslant 0$.

3. Consider the Burger's equation $u_t = uu_x + u_{xx}$. Show that it can be linearized by means of a Riccati type transformation and thus solved in terms of solutions of the heat equation.

4. Consider the recurrence relation

$$u(x, t+\Delta) = au(x + bu\delta, t) + (1-a)\left[\frac{u(x+\delta, t) + u(x-\delta, t)}{2}\right].$$

Show that the parameters a, b, δ can be chosen subject to $0 < a < 1$, $b > 0$, $\delta > 0$ so that the relation is consistent to order Δ with the Burger's equation. Show that nonnegativity and boundedness are preserved.

5. Consider the system

$$u_t = uu_x + au_{xx}, \qquad u(x, 0) = g(x), \tag{1}$$

show that we can write

$$\begin{aligned} u(x, t+\Delta) &= \alpha u(x - au(x,t)\Delta, t) + \frac{1-\alpha}{2}[u(x + b\Delta^{1/2}, t) + u(x - b\Delta^{1/2}, t)] \\ &\quad - 3\beta\left[h\left(x - \frac{\Delta}{\beta}h(x,t), t\right) - h(x,t)\right], \end{aligned} \tag{2}$$

$$\begin{aligned} h(x, t+\Delta) &= \gamma h(x - cu(x,t)\Delta, t) + \frac{1-\gamma}{2}[h(x + d\Delta^{1/2}, t) + h(x - d\Delta^{1/2}, t)] \\ &\quad - \delta\left[u\left(x - \frac{\Delta}{\delta}h(x,t), t\right) - u(x,t)\right], \end{aligned}$$

where Δ is the integration step size and α, β, γ, δ, a, b, c and d are constants which will be determined. Obtain a recurrence relation valid to $O(\Delta)$ if

$$a = 1/\alpha, \qquad b = [2/(1-\alpha)\, Re]^{1/2}, \qquad c = 1/\gamma, \qquad d = [2/(1-\gamma)\, Rm]^{1/2}. \tag{3}$$

Dah-Teng Jong, "Direct Computational Approaches to a Magnetohydradynamic Model System," (to appear).

7. Differential Quadrature

Let us finally recall the method of differential quadrature of an earlier chapter. Let $x_1, x_2, \ldots, x_N$ be a set of points in $[0, 1]$ and write

$$u_x\big|_{x=x_i} \cong \sum_{j=1}^{N} a_{ij} u(x_j, t) \tag{1}$$

where the a_{ij} are chosen in some convenient fashion. Then the partial differential equation

$$u_t = uu_x, \qquad u(x,0) = g(x), \tag{2}$$

is replaced by the system of ordinary differential equations

$$dv_i/dt \cong v_i \sum_{j=i}^{N} a_{ij} v_j, \qquad v_i(0) = g(x), \quad i = 1, \ldots, N, \tag{3}$$

where $v_i \cong u(x_i, t)$.

BIBLIOGRAPHY AND COMMENT

Section 4. See

R. Bellman, I. Cherry, and G. M. Wing, *Quart. Appl. Math.*, Vol. 48, 1962, p. 1325.

S. P. Azen, R. Bellman, and J. M. Richardson, *Another Computational Approach to A Mathematical Model of Turbulence*, RM-3918-ARPA, The RAND Corporation, 1964.

Section 5. See

R. Bellman and K. L. Cooke, "Existence and Uniqueness Theorems," in *Invariant Imbedding–II: Convergence of a New Difference Algorithm, J. Math. Anal. Appl.*, Vol. 12, 1965, pp. 247–253.

Section 6. See

S. Azen, *Higher Order Approximations to the Computational Solution of Partial Differential Equations*, RM-3917-ARPA, The RAND Corporation, 1964.

Section 7. See

R. Bellman, J. Casti, and B. Kashef, "Differential Quadrature, A Rapid Method for the Computational Solution of Nonlinear Partial Differential Equations," (to appear).

Chapter 10

Parabolic Equations

1. The Heat Equation

Perhaps the most important example of a parabolic partial differential equation is the heat equation. For a uniform thin rod, the temperature distribution is given by the solution of

$$u_t = u_{xx} , \tag{1}$$

where the initial distribution

$$u(x,0) = g(x) , \tag{2}$$

is known and we also know the conditions at both ends of the rod, i.e.,

$$u(0,t) = a(t) , \qquad u(1,t) = b(t) . \tag{3}$$

We will be concerned with the more general problem of heat flow over a region R with boundary Γ. Thus, we will consider the solution of

$$u_t = k(x,y)\,[u_{xx} + u_{yy}] , \tag{4}$$

where

$$k(x,y) > 0 , \tag{5}$$

in R,

$$u(x,y,0) = g(x,y)\,, \tag{6}$$

and

$$u(x,y,t) = f(x,y)\,, \tag{7}$$

on Γ, a combination of boundary and initial conditions. We will see that to obtain heat distributions for a one-dimensional rod, we can apply scalar versions of the methods we have developed for the two-space-dimension problem.

A basic approach that can be employed is once again a finite difference method. Unfortunately we immediately encounter two new difficulties; dimensionality and stability. The dimensionality difficulty arises since we are now considering an equation with three independent variables rather than two. We must be careful or the number of operations can easily get out of hand. We will find, unfortunately, that the simplest computational schemes possess certain unfavorable stability properties. Before we consider the actual methods, let us discuss some mathematical difficulties which we have thus far glossed over.

EXERCISES

1. Let u satisfy $u_t = u_{xx}$, $u(x,0) = g(x)$, $u(0,t) = u(1,t) = 0$.
Show that

$$(d/dt)\left(\int_0^1 u^2\,dx\right) = 2\int_0^1 uu_t\,dx = 2\int_0^1 uu_{xx}\,dx = -2\int_0^1 u_x^{\,2}\,dx\,.$$

Hence show that

$$\int_0^1 u^2\,dx \leqslant \int_0^1 g^2(x)\,dx\,.$$

2. Show similarly that

$$\int_0^1 u^{2n}\,dx$$

is decreasing as a function of t for $t \geqslant 0$.

3. Show that

$$\max_x |u| = \lim_{n\to\infty}\left(\int_0^1 u^{2n}\,dx\right)^{1/2n}.$$

4. Hence show that max x $|u|$ is decreasing as a function of t.

5. Show that for some $\lambda > 0$.

$$\int_0^1 u^2\,dx \leqslant \left(\int_0^1 g^2(x)\,dx\right) e^{-\lambda t}.$$

6. Derive similar results for the equation $u_t = uu_x + u_{xx}$, $u(0,t) = u(1,t) = 0$.

7. Derive similar results for the equation $u_t = u_{xx} + u_{yy}$, $u = 0$ on Γ, $t \geqslant 0$.

2. Properly Posed Problems

In our previous work we have tacitly made a number of assumptions. Perhaps the most basic assumption we have made is that the finite difference techniques will yield an approximate solution to the problem of interest. As an example of a problem where approximate techniques are not valid, consider the motion of a pendulum on the end of a rigid rod as shown in Fig. 1. The motion of the pendulum is described by the equation

$$d^2\theta/dt^2 + k\sin\theta = 0\,. \tag{1}$$

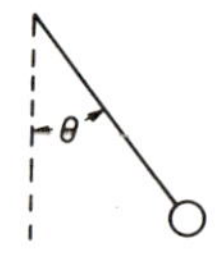

FIGURE 1

Suppose we place the pendulum in an upright position, i.e.,

$$\theta(0) = \pi\,, \qquad \theta'(0) = 0\,. \tag{2}$$

Then (1) and (2) have the solution

$$\theta(t) = \pi\,, \qquad t > 0\,. \tag{3}$$

Suppose, however, we were to attempt the solution of (1) and (2) by digital computer. Since π is irrational we would actually start with the initial conditions

$$\theta(0) = \pi - \varepsilon\,, \qquad \theta'(0) = 0\,, \tag{4}$$

$\varepsilon \ll 1$. Unfortunately (1) and (4) possess an oscillatory solution with $\theta(t)$ going from $\pi - \varepsilon$ to $-\pi + \varepsilon$. This is an example of an ill-posed or improperly posed problem. The investigation of such problems goes back to Hadamard.

We will always assume that the problems we consider are properly posed (or well-posed). A problem will be called *properly posed* if the solution exists, is unique and depends continuously on the input data. Thus, the above example was not properly posed since a small change in the input data, the initial conditions, did not produce a correspondingly small change in the solution.

3. Consistency and Stability

Our numerical technique has been to replace the boundary value problem of interest by a set of difference equations. We have assumed that as the grid spacing approached zero the solution of the difference equations converged uniformly to the solution of the problem of interest. For the problems we have considered so far it is a simple matter to prove that our methods converge in the above sense. However, for parabolic equations which are initial value problems in the time variable, the notion of convergence is a bit more complex. It is tied up with the problems of consistency and stability.

To replace parabolic equations by difference equations we will have to consider both discrete values of the time variable and discrete values of the space variables. Let us denote the time and space grid spacings by Δ and h respectively. We will call a set of difference equations *consistent* with a given partial differential equation if in the limit as Δ and h go to zero independently, the difference equations yield the desired partial differential equation. There are many examples in the literature of methods which are consistent only if Δ and h go to zero in a particular way. We will only consider discretizations which are consistent with the parabolic equations with which we are dealing.

We will call a difference approximation *stable* if in the limit as the grid spacing goes to zero the effect of any piece of input data or the effect of an error in the computation does not yield unbounded contributions to the solution. We can understand what this concept of stability means on physical grounds. Since we are interested in parabolic (or diffusion) equations we know that the given partial differential equation can have no unbounded solutions. Thus if the difference approximation is to model the partial differential equation of interest, we must require that our difference approximation does not introduce unbounded solutions which cannot exist physically. This new definition of stability includes the concept of computational stability introduced previously.

The relation between convergence, consistency and stability is given by a theorem due to P. D. Lax which we state without proof.

Theorem. *Given a properly posed initial value problem and a finite difference approximation to it that satisfies the consistency condition, stability is a necessary and sufficient condition for convergence.*

Since we have agreed to consider only consistent methods, we will investigate only the stability of the various approximations.

EXERCISES

1. Let $u_{i,j}$ denote $u(ih, j\Delta)$. Show that as h and $\Delta \to 0$, the difference equations

$$(u_{i,j+1} - u_{i,j})/\Delta = (u_{i+1,j} - 2u_{i,j} + u_{i-1,j})/h^2 \tag{1}$$

yield the heat equation

$$u_t = u_{xx} . \tag{2}$$

Thus, (1) is consistent with (2).

2. Show that the Dufort and Frankel scheme

$$(u_{i,j+1} - u_{i,j-1})/2\Delta = (u_{i-1,j} - u_{i,j-1} - u_{i,j+1} + u_{i+1,j})/h^2 \tag{3}$$

is consistent with (2) if Δ and $h \to 0$ is such a way that Δ/h^2 is constant. Show that if Δ and $h \to 0$ in such a manner that $\Delta/h = c$, then (3) is consistent with the equation

$$u_{xx} = u_t + c^2 u_{tt} .$$

4. Explicit Methods

We can now consider some typical difference schemes. Since we would like to analyze the stability properties of the methods in some detail, we will use as an example the heat equation

$$u_t = c[u_{xx} + u_{yy}] , \tag{1}$$

where c is a positive constant and the region of interest is the unit square. We assume that we are given $u(x, y, 0)$ and also the values on the boundaries of the region for all t. We will use the following discretization

$$x_i = ih , \qquad y_j = jh , \qquad t_l = l\Delta , \tag{2}$$

where

$$Nh = 1 . \tag{3}$$

We will let $u_{ij}(t_l)$ denote $u(ih, jh, l\Delta)$. To obtain a difference equation corresponding to (1) we approximate u_t by

$$u_t(x,y,t) = \frac{u(x,y,t+\Delta) - u(x,y,t)}{\Delta} + O(\Delta) \tag{4}$$

and $u_{xx}+u_{yy}$ by

$$u_{xx} + u_{yy} = \frac{u(x+h,y,t) + u(x,y+h,t) - 4u(x,y,t)}{h^2} + \frac{u(x-h,y,t) + u(x,y-h,t)}{h^2} + O(h^2). \tag{5}$$

Thus at $x = ih$, $y = jh$, $t = l\Delta$ we have

$$\frac{u_{ij}(t_{l+1}) - u_{ij}(t_l)}{\Delta} = \frac{c}{h^2}\,[u_{i+1,j}(t_l) + u_{i,j+1}(t_l) - 4u_{ij}(t_l) + u_{i-1,j}(t_l) - u_{i,j-1}(t_l)] + O(\Delta + O(h^2)). \tag{6}$$

Rearranging (6) yields the difference scheme

$$u_{ij}(t_{l+1}) = u_{ij}(t_l) + (c\Delta/h^2)\,[u_{i+1,j}(t_l) + u_{i,j+1}(t_l) - 4u_{ij}(t_l) + u_{i,j-1}(t_l)]. \tag{7}$$

Since we are given $\{u_{ij}(t_0)\}$ and the values $u_{ij}(t_l)$ for i and j equal to 0 and N, (7) can be solved in a trivial manner. Each $u_{ij}(t_{l+1})$ can be found independently of the others; thus the method is explicit. Unfortunately, as we shall now see, this method is unstable unless the grid spacing h is made sufficiently small.

Let the matrix Q be defined as before

$$Q = (q_{ij}), \qquad q_{ij} = \begin{cases} 2, & i = j, \\ -1, & |i-j| = 1, \\ 0, & \text{otherwise}, \end{cases} \tag{8}$$

and the matrix U_l will be defined as

$$U_l = (u_{ij}(t_l)), \qquad i,j = 1, \ldots, N-1. \tag{9}$$

Using this notation (7) can be written as

$$U_{l+1} = U_l - r[QU_l + U_l Q] + S_l, \tag{10}$$

where the matrix S_l is determined by the boundary conditions and we have introduced the ratio $r = c\Delta/h^2$. In order to investigate the stability of this explicit difference scheme we will examine under what circumstances the homogeneous form of (10)

$$U_{l+1} = U_l - r[QU_l + U_l Q]\,, \tag{11}$$

has no unbounded solutions. We are justified in examining the homogeneous form only since if we were to perturb (10) we would see that the error would satisfy (11). This is equivalent to the error analysis techniques we used in Chapters 5 and 6.

Let u_l be the vector of dimension $(N-1)^2$ obtained by taking the elements of U_l by columns. As we did in Chapter 8 using Kronecker products we can replace (11) by

$$u_{l+1} = [I - r(I \otimes Q + Q \otimes I]u_l\,. \tag{12}$$

For (12) to have no unbounded solutions we must have

$$\rho[I - r(I \otimes Q + Q \otimes I)] < 1\,. \tag{13}$$

If μ_i is the ith characteristic value of Q, then the characteristic values of $I - r(I \otimes Q + xI)$ are the numbers

$$\lambda_k = 1 - r(\mu_i + \mu_j)\,. \tag{14}$$

Since we have already shown that

$$0 < \mu_i < 4\,, \tag{15}$$

then (13) will be satified if

$$r < \tfrac{1}{4}\,. \tag{16}$$

We have seen that as h gets small the largest characteristic value of Q gets arbitrarily close to 4. Thus the bound in (16) is a very good one and we see what a severe restriction (16) places on the method. We note that for

$$c = 1\,, \qquad h = 0.1\,, \tag{17}$$

we must choose a time step Δ such that

$$\Delta < 0.0025\,. \tag{18}$$

Thus, the simplicity of this method is counteracted by the necessity of taking so many time steps that the amount of computation becomes unreasonable.

EXERCISE

Show that for the one space variable heat equation

$$u_t = cu_{xx}$$

the explicit method requires that $r < 1$ for stability.

5. Implicit Methods

Instead of the foward difference formula, (4.4), we used to approximate u_t we will now use the backward difference

$$u_t(x,y,t) = \frac{u(x,y,t) - u(x,y,t-\Delta)}{\Delta} + O(\Delta), \tag{1}$$

which has the same truncation error. Combining (1) and (4.5) we have the difference equations

$$\begin{aligned} u_{ij}(t_l) - u_{ij}(t_{l-1}) = r[&u_{i+i,j}(t_l) + u_{i,j-1}(t_l) - 4u_{ij}(t_l) \\ &+ u_{i-1,j} + u_{i,j-1}(t_l)] . \end{aligned} \tag{2}$$

We see immediately that this difference scheme is implicit. At each stage we have the values $\{u_{ij}(t_{l-1})\}$ and must then solve a set of algebraic equations for the values $\{u_{ij}(t_l)\}$. In fact the solution of (2) is equivalent to solving an elliptic equation of the type we have been considering. In order to clarify this point let us return to the original parabolic equation

$$u_t = c(u_{xx} + u_{tt}) . \tag{3}$$

Using the approximation in (1) to replace u_t we have

$$u(x,y,t) - u(x,y,t-\Delta) = c\Delta[u_{xx}(x,y,t) + u_{yy}(x,y,t)) , \tag{4}$$

an elliptic equation for any fixed value of t. If we treat (4) by the methods we have developed for elliptic equations, we will immediately arrive at (2). We see that implicit methods are equivalent to solving an elliptic equation at each time step and thus can require a large amount of computation.

The stability of (2) can be investigated directly. If U_l, Q and S_l are defined as before, (2) can be written as

$$U_l - U_{l-1} = -r[QU_l + U_l Q] + S_l . \tag{5}$$

Again we consider the homogeneous matrix vector version of (5)

$$u_l - u_{l-1} = -r[Q \otimes I + I \otimes Q] u_l , \tag{6}$$

or

$$u_l = [I + r(Q \otimes I + I \otimes Q)]^{-1} u_{l-1} . \tag{7}$$

If μ_i is a characteristic value of Q then the characteristic values of $[I+r(Q\times I+I\times Q]^{-1}$ are the numbers

$$\lambda_k = 1/[1 + r(\mu_i + \mu_i)] . \tag{8}$$

Hence

$$0 < \lambda_k < 1 \tag{9}$$

and (5) can have no unbounded solutions. Consequently, the implicit method is stable for all values of h and Δ.

6. Crank–Nicolson Method

We have seen that the simple implicit method we have just derived is always stable but requires the solution of an elliptic equation at each time stage. Since the truncation error was $O(\Delta)+O(h^2)$, we have to use a very very small time step to achieve reasonable accuracy. Thus, this simple method is usually impractical since it will require a great deal of computation. We can get around this difficulty by using more accurate implicit difference schemes.

The Crank–Nicolson method is derived by approximating the partial derivatives around $u(x, y, t+\Delta/2)$. It is easy to check by means of the Taylor series that u_t can be written as

$$u_t(x, y, t+\Delta/2) = \frac{u(x, y, t+\Delta) - u(x, y, t)}{\Delta} + O(\Delta^2) \tag{1}$$

For $u_{xx}+u_{yy}$ we use

$$\begin{aligned} u_{xx}(x, y, t + \Delta/2) + u_{yy}(x, y, t + \Delta/2) = {} & \tfrac{1}{2}[u_{xx}(u, y, t + \Delta) + u_{yy}(x, y, t + \Delta) \\ & + u_{xx}(x, y, t) + u_{yy}(x, y, t)] \\ & + O(\Delta^2) . \end{aligned} \tag{2}$$

Now we can use the standard approximation, (4.5), to replace $u_{xx}+u_{yy}$ on the right of (2). Combining (1), (2) and (4.5) we have the Crank–Nicolson method

$$\begin{aligned} u_{ij}(t_{l+1}) - u_{ij}(t_l) = {} & (r/2)[u_{i+1,j}(t_{l+1}) + u_{i,j-1}(t_{l+1}) - 4u_{ij}(t_{l+1}) \\ & + u_{i-1,j}(t_{l+1}) + u_{i,j-1}(t_{l+1}) + u_{i+1,j}(t_l) + u_{i,j+1}(t_l) \\ & - 4u_{ij}(t_l) + u_{i-1,j}(t_l) + u_{i,j-1}(t_l)] , \end{aligned} \tag{3}$$

which is correct $O(\Delta^2) + O(h^2)$. We can easily see by using (1) and (2) in the original parabolic equation that (3) is the finite difference approximation of the elliptic equation

$$u_{xx}(x,y,t+\Delta) + u_{yy}(x,y,t+\Delta) - (2\Delta/c)\,u(x,y,t+\Delta)$$
$$= -u_{xx}(x,y,t) - u_{yy}(x,y,t) + (2\Delta/c)\,u(x,y,t). \quad (4)$$

Thus the Crank–Nicolson method requires essentially the same amount of computation per time step as the implicit method of Section 5. Now, however, we can take larger time steps since the truncation error is smaller.

Let us now show that the Crank–Nicolson method is stable for all positive step sizes. Using our previous definitions, (3) can be written as

$$U_{l+1} - U_l = -r[U_{l+1}Q + QU_{l+1} + U_l Q + QU_l] + S_l \quad (5)$$

and the homogeneous matrix vector form is

$$u_{l+1} - u_l = -(r/2)[(I\otimes Q + Q\otimes I)u_{l+1} + (I\otimes Q + Q\otimes I)u_l]. \quad (6)$$

Solving for u_{l+1}, we find

$$u_{l+1} = [I + (r/2)(I\otimes Q + Q\otimes I)]^{-1}[I - (r/2)(I\otimes Q + Q\otimes I)]\,u_l. \quad (7)$$

Thus, the significant characteristic values are the numbers

$$\lambda_k = \frac{1 - (r/2)(\mu_i + \mu_j)}{1 + (r/2)(\mu_i + \mu_j)}, \quad (8)$$

where μ_i is a characteristic value of Q. Once again since

$$0 < \mu_i < 4, \quad (9)$$

we have

$$0 < \lambda_k < 1, \quad (10)$$

and (3) is stable for all positive values of Δ and h.

7. Alternating-Direction Implicit Methods

For a fixed value of t let us consider the following equations

$$\frac{v(x,y) - u(x,y,t)}{\Delta/2} = c[u_{xx}(x,y,t) + v_{yy}(x,y)] \quad (1)$$

and

$$\frac{w(x,y) - v(x,y)}{\Delta/2} = c[w_{xx}(x,y) + v_{yy}(x,y)]. \tag{2}$$

If we assume that $u(x,y,t)$ is known then we can regard (1) as an equation for $v(x,y)$ and (2) as an equation for $w(x,y)$ using the solution of (1). Using Taylor series expansions, it is not difficult to show* that if $u(x,y,t)$ is the solution of

$$u_t = c(u_{xx} + u_{yy}) \tag{3}$$

then

$$w(x,y) = u(x,y,t+\Delta) + O(\Delta^2). \tag{4}$$

Thus we expect this method to be as accurate as the Crank–Nicolson method. However, the computational advantages of this approach are great. Equation (1) involves only y-derivatives of the unknown function v while (2) contains only x-derivatives of the unknown function w; hence the name alternating-direction implicit method. At each stage of the computation we solve two much simpler equations rather than solving an elliptic equation such as (5.4) or (6.4). We will now examine the difference equations corresponding to (1) and (2) to see how simple these computations are.

Let $u_{ij}(t_{l+(1/2)})$ denote the finite difference approximation of v. Although the values $u_{ij}(t_{l+(1/2)})$ arc not directly related to the values of u at $t+\Delta/2$, this notation is standard. Then using our standard approximations for second derivatives (1) and (2) become

$$\begin{aligned} u_{ij}(t_{l+1/2}) - u_{ij}(t_l) = (r/2)[&u_{i+1,j}(t_l) - 2u_{ij}(t_l) + u_{i-1,j}(t_l) \\ &+ u_{,j+1}(t_{l+1/2}) - 2u_{ij}(t_{l+1/2}) + u_{i,j-1}(t_l)], \end{aligned} \tag{5}$$

and

$$\begin{aligned} u_{ij}(t_{l+1}) - u_{ij}(t_{l-1/2}) = [&u_{i+1,j}(t_{l+1}) - u_{ij}(t_{l+1}) + u_{i+1,j}(t_{l+1}) \\ &+ u_{i,j+1}(t_{l+1/2}) - 2u_{ij}(t_{l+1/2}) + u_{i,j-1}(t_{l+1/2})]. \end{aligned} \tag{6}$$

We can show that the truncation error is $O(\Delta^2)+O(h^2)$ directly from (5) and (6). Converting (5) and (6) to matrix form we have

$$[I + (r/2)Q]\,U_{l+1/2} = U_l[I - (r/2)Q] + S_l, \tag{7}$$

---

* We assume that the appropriate boundary conditions are used in the solution of (1) and (2).

and

$$U_{l+1}[I+(r/2)Q] = [I-(r/2)Q]U_{l+1/2} + S_l. \tag{8}$$

Since Q is tridiagonal, each of these systems can be solved in a simple manner. As we saw in Chapter 8, we will have scalar versions of the matrix–vector computations necessary to solve elliptic equations.

We can analyze the stability of this method by considering the homogeneous version of (7) and (8). Combining the two equations we have

$$U_{l+1} = [I-(r/2)Q][I+(r/2)Q]^{-1}U_l[I-(r/2)Q][I+(r/2)Q]^{-1}. \tag{9}$$

It is clear that the matrices $I-(r/2)Q$ and $I+(r/2)Q$ commute. If we let T denote the matrix

$$T = [I-(r/2)Q][I+(r/2)Q]^{-1}, \tag{10}$$

then from (9) we have

$$U_{l+1} = T^l U_0 T^l. \tag{11}$$

Since the characteristic values of T are the numbers

$$\lambda_i = \frac{1-(r/2)\mu_i}{1+(r/2)\mu_i}, \tag{12}$$

where μ_i is a characteristic value of Q, we have

$$0 < |\lambda_i| < 1, \tag{13}$$

and the method is stable for all positive Δ and h.

Since the potential equation can be looked at as the steady state solution of

$$u_t = c[u_{xx} + u_{yy}], \tag{14}$$

the high efficiency of the alternating direction implict method suggests that instead of solving the potential equation we solve (14) until steady state is reached. In fact, it is precisely in this manner that the alternating direction implicit method for elliptic equations was originally derived. We will not pursue the investigation of the alternating direction implicit method any further, since the analysis can become very complex and the methods so far have limited applicability.

8. The Laplace Transform

In this section we will present an entirely different approach to the solution of parabolic equations, based upon the use of the Laplace transform. The Laplace transform is a fundamental method by which we can reduce the level of transcendence of a large number of equations. We will assume that the reader has some familiarity with the Laplace transform.

Let us recall that the Laplace transform, $L(u)$, of the function $u(t)$ is defined as

$$L(u) = F(s) = \int_0^\infty u(t)\, e^{-st}\, dt\,. \tag{1}$$

We will assume that the integral converges absolutely for $s \geqslant 0$. It is easy to show that

$$L(du/dt) = sF(s) - u(0)\,. \tag{2}$$

Therefore the Laplace transform can be used to reduce ordinary linear differential equations with constant coefficients to algebraic equations.

Returning to our original equation

$$\partial u/\partial t = c[\partial^2 u/\partial x^2) + (\partial^2 u/\partial y^2)]\,, \tag{3}$$

where

$$u(x, y, 0) = g(x, y)\,, \tag{4}$$

and $u(x, y, t)$ is specified on the edges of the unit square, we are led to transform this equation. We find by applying (1) and (2) and defining

$$\tilde{u}(x, y, s) = L(u(x, y, t))\,, \tag{5}$$

that $\tilde{u}$ satisfies

$$s\tilde{u} - g(x, y) = k(x, y)[(\partial^2 \tilde{u}/\partial x^2) + (\partial^2 \tilde{u}/\partial y^2)]\,, \tag{6}$$

an elliptic equation for any given value of s. The boundary conditions for (6) are simply the transforms of the original boundary conditions. If the boundary conditions on the edges of the square are independent of t, then the boundary conditions for (6) are clearly just the original boundary conditions divided by s.

Our approach will be to solve (6) for given values of s, such as $1, 2, \ldots, N$, by the methods we have thus far developed. Then we will have a table of particular values of the Laplace transform of the function we seek. The success of this procedure rests on our ability to invert the Laplace transform numerically. If we can conveniently invert the transform, then we can hope to solve the original parabolic equation by solving a relatively small number of elliptic equations.

9. Gaussian Quadrature

The numerical inversion of the Laplace transform will be based upon the use of a quadrature formula for the integral which defines the Laplace transform. Let us then first review a few basic facts about Gaussian quadrature. Suppose we wish to approximate the definite integral

$$I = \int_0^1 f(x)\,dx\,, \tag{1}$$

by a finite sum

$$I_s = \sum_{i=1}^{N} w_i f(x_i)\,. \tag{2}$$

The $\{x_i\}$ are known as the quadrature points and $\{w_i\}$ are a set of weights. We are required to pick the quadrature points and weights so that (1) and (2) agree whenever $f(x)$ is an arbitrary polynomial of degree up to $2N-1$. Let us not go through the details of how the points and weights are determined since the derivation can be found in any standard text on numerical analysis. We will merely quote the basic result.

If $P_N(x)$ denotes the shifted Legendre polynomial of order N then the $\{x_i\}$ are the zeros of this polynomial, all of which are distinct and lie between zero and one. If we define functions $L_i(x)$ by

$$L_i(x) = \prod_{\substack{j=1 \\ j\neq 1}}^{N} (x - x_j) = P_N(x)/(x - x_i)\,, \tag{3}$$

then the weights can be determined by the expressions

$$w_i = [1/L_i(x_i)] \int_0^1 L_i(x)\,dx\,, \qquad i = 1, 2, \ldots, N\,. \tag{4}$$

Both the zeros and the weights can either be found in tables or computed as needed to an arbitrary degree of accuracy. The accuracy of the quadrature formula depends upon how well the function $f(x)$ can be approximated by a polynomial. We will return to this point in a moment.

If we make the transformation

$$x = e^{-t} \tag{5}$$

in the definition of the Laplace transform

$$F(s) = \int_0^\infty u(t)\, e^{-st}\, dt \tag{6}$$

we find

$$F(s) = \int_0^1 u(-\log x)\, x^{s-1}\, dx\,. \tag{7}$$

This will be our starting point. In view of (5) the requirement that the quadrature formula is a good approximation to (7) is that $u(t)$ can be well approximated by the finite sum of exponentials, i.e.,

$$u(t) \cong \sum_{i=1}^{N} a_k e^{-kt}\,. \tag{8}$$

We will assume that this is always the case.

10. Inversion of the Laplace Transform

We define

$$g(x) = u(-\log x)\,, \tag{1}$$

and rewrite (9.7) as

$$F(s) = \int_0^1 x^{s-1} g(x)\, dx\,. \tag{2}$$

Applying the Gaussian quadrature formula of order N to (2) we obtain the approximate formula

$$F(s) = \sum_{i=1}^{N} w_i x_i^{s-1} g(x_i)\,. \tag{3}$$

If we can find $F(s)$ numerically for $s = 1, 2, \ldots, N$, then (3) yields the set of N linear algebraic equations

$$\sum_{i=1}^{N} w_i x_i^k g(x_i) = F(k+1)\,, \qquad k = 0, 1, \ldots, N-1\,, \tag{4}$$

in the unknown $g(x_i)$. To simplify the above expression we define

$$y_i = w_i g(x_i)\,, \qquad a_k = F(k+1)\,, \tag{5}$$

and obtain the simpler equations

$$\sum_{i=1}^{N} x_{ik} y_i = a_k\,. \tag{6}$$

This is equivalent to the matrix–vector equation

$$Xy = a\,, \tag{7}$$

with

$$X = (x_i^{j-1})\,. \tag{8}$$

The matrix X is obviously the Vandermonde matrix which, although nonsingular since the x_i are distinct, is highly ill-conditioned. Thus, while the system (6) is completely determined it cannot readily be solved numerically if $N \gg 1$. However, if we proceed a bit further we can obtain an explicit inversion formula which can be computed to a high degree of accuracy.

We start by multiplying the kth equation of (6) by an unspecified parameter q_k. Adding the N equations we obtain

$$\sum_{i=1}^{N} y_i \left(\sum_{k=0}^{N-1} q_k x_{ik} \right) = \sum_{k=0}^{N-1} a_k q_k\,. \tag{9}$$

Setting

$$f(x) = \sum_{k=0}^{N-1} q_k x^k\,, \tag{10}$$

we have

$$\sum_{i=1}^{N} y_i f(x_i) = \sum_{k=0}^{N-1} q_k a_k\,. \tag{11}$$

The polynomial of degree $N-1$, $f(x)$ is still unspecified. Let us, however, repeat the above procedure N times and derive N distinct polynomials $f(x)$ and consequently N different sets of coefficients q_k. Thus we set

$$f(x) = f_j(x)\,, \qquad q_k = q_{kj}\,, \qquad j = 1, 2, \ldots, N\,. \tag{12}$$

Since we are interested in finding the y_i, we choose $f_j(x)$ so that

$$\begin{aligned} f_j(x_i) &= 0\,, \qquad i \neq j\,, \\ f_i(x_i) &= 1\,, \qquad i = j\,. \end{aligned} \tag{13}$$

With this convenient choice of the $f_j(x)$, (11) reduces to

$$y_i = \sum_{k=0}^{N-1} a_k q_{ki}\,. \tag{14}$$

If we can determine the $f_j(x)$, and thus obtain the coefficients q_{ki} numerically, we will have our inversion formula. Since the x_i are the roots of the shifted

Legendre polynomial of order N, $P_N(x)$, we can write $P_N(x)$ as

$$P_N(x) = c\prod_{i=1}^{N}(x - x_i). \tag{15}$$

The first condition of (13) requires that

$$f_j(x) = P_N(x)/[(x - x_j)P_{N'}(x_j)]. \tag{16}$$

The coefficients q_{jk} are the coefficients of this polynomial. Since the zeros of $P_N(x)$ are available and the coefficients of $P_N(x)$ are integers, the q_{jk} can be determined to any desired degree of accuracy.

If we use the substitutions of (1) and (5) in (14) we can rewrite our inversion formula as

$$u(-\log x_i) = (1/w_i)\sum_{k=0}^{N-1} q_{ik}F(k+1). \tag{17}$$

11. Computational Aspects

The coefficients q_{jk}/w_k have been computed for values of N up to 15 and are given in the reference cited at the end of the chapter. The coefficients can easily be computed using a short subroutine whose input is the set of zeros of the Legendre polynomial and the associated weights. This subroutine can also be found in the cited reference. We elected to use the second method

The condition that

$$u(t) \cong \sum_{i=1}^{N} a_k e^{-kt} \tag{1}$$

is essentially a "smoothness" requirement. Thus, we do not consider equations whose solutions possess any kind of spike or high frequency behavior. We cannot hope to have any one method which would be suitable for all equations since we know that the Laplace inverse is an unbounded operator. Variants of the method can be devised for solutions with singular behavior.

We are given the values of the desired function at the points

$$t_i = -\log x_i, \tag{2}$$

where the x_i are the zeros of the Legendre polynomial. Since the x_i are uniformly distributed on [0, 1], the t_i are clustered near zero. In most problems this will not be a significant factor since we can use a suitable interpolation formula to find $u(t)$ at any desired t. We can also use some of the properties of the Laplace transform to given values of $u(t)$ at points other than those in (2).

BIBLIOGRAPHY AND COMMENT

Sections 1–3. See

R. D. Richtmeyer and K. W. Morton, *Difference Methods for Initial Value Problems*, Wiley (Interscience), New York, 1967.

P. Lax, "Numerical Solutions of Partial Differential Equation," *Amer. Math. Monthly*, Vol. 72, 1965, pp. 74–84.

E. Isaacson and H. B. Keller, *Analysis of Numerical Methods*, Wiley, New York, 1966.

Section 4. In addition to the above references see the chapter by H. B. Keller in

A. Ralston and H. S. Wilf, eds., *Mathematical Methods for Digital Computer*, Wiley, New York, 1960.

Also see the survey article and references by D. Young in

J. Todd, ed., *Survey of Numerical Analysis*, McGraw-Hill, New York, 1962.

Section 6. See

J. Crank and P. Nicolson, "A Practical Method for Numerical Evaluation of Solutions of Partial Differential Equations of the Heat Conduction Type," *Proc. Cambridge. Philos. Soc.* Vol. 43, 1947, pp. 50–67.

Section 7. See

D. W. Peaceman and H. H. Rachford Jr., "The Numerical Solution of Parabolic and Elliptic Differential Equations," *J. Soc. Indust. Appl. Math.*, Vol 3, 1955, pp. 28–41.

Section 8. See

R. Bellman, R. Kalaba, and J. Lockett, *Numerical Inversion of the Laplace Transform*, Amer. Elsevier, New York, 1966.

E. Angel, *Numerical Inversion of the Laplace Transform and Multidimensional Heat Equations*, Electronic Sciences Laboratory Rept. 70-5, Univ. of Southern California, 1970.

B. Van der Pol and H. Bremmer, *Operational Calculus Based on the Two-Sided Laplace Integral*, Cambridge Univ. Press, London and New York, 1955.

Section 9. See

C. Lanczos, *Applied Analysis*, Prentice-Hall, Englewood Cliffs, New Jersey, 1956.

Section 10. For a proof of the nonsingularity of the Vandermonde matrix see

R. Bellman, *Introduction to Matrix Analysis*, 2nd ed., McGraw-Hill, New York, 1970.

Section 11. The coefficients appear in the book by Bellman, Kalaba, and Lockett cited under Section 8.

Chapter 11

Nonlinear Equations and Quasilinearization

1. Introduction

In the foregoing chapters we have considered linear partial differential equations exclusively. The linearity was essential for the invariant imbedding approach while the quadratic nature of the associated variational problems was equally essential for the simplicity of the final results obtained using dynamic programming. In this chapter we wish to extend the scope of our methods to treat nonlinear equations such as

$$u_{xx} + u_{yy} = e^u, \tag{1}$$

an equation arising in magnetohydrodynamics and elsewhere, and the nonlinear heat equation

$$v_t = v_{xx} + v_{yy} + g(v). \tag{2}$$

The most direct method of obtaining a numerical solution is by way of a discretization of the equation. For reasons we have already mentioned in the linear case this approach presents many serious obstacles that are greatly magnified by the nonlinearity. We shall therefore pursue a different route here.

At the end of the chapter we apply the methods briefly to some identification problems.

2. Successive Approximations

We have previously demonstrated that we can develop a simple computer program which will yield the numerical solution of the general linear equation

$$u_{xx} + u_{yy} + h(x,y)u = f(x,y) \tag{1}$$

over a region of very general type. Our aim then is to solve an equation such as (1) as the limit of the solutions of a sequence of linear equations of the foregoing form. This involves the method of successive approximations, the general factotum of analysis.

The simplest variant of the method is that of Picard. In place of (1) we consider the sequence of *linear* equations

$$u_{xx}^{(n+1)} + u_{yy}^{(n+1)} = \exp(u^{(n)}), \tag{2}$$

where $u^{(0)}$ is specified, the initial guess. Similarly, in place of (1.2) we consider the sequence of linear equations

$$v_t^{(n+1)} = v_{xx}^{(n+1)} + v_{yy}^{(n+1)} + g(v^{(n)}), \tag{3}$$

again with $v^{(0)}$ specified.

Presumably, this resolves the matter. There are, however, two major sources of difficulty. In the first place the convergence is slow, geometric convergence at best. In the second place, unless $u^{(0)}$ and $v^{(0)}$ are carefully chosen there is no convergence at all.

For these reasons we will employ a different variant of successive approximations, one generated by considerations of the theory of dynamic programming.

3. Quasilinearization

A fundamental class of equations in the theory of dynamic programming has the form

$$\max_v T(u,v) = 0, \tag{1}$$

where T is linear in u. The function u represents the return function, while v represents the policy function. In solving (1) we can employ approximation in function space, the classical approach, or approximation in policy space, a new approach, with a number of desirable features.

If we guess a policy, v_0, the equation for the associated initial return function u_0 is linear,

$$T(u_0, v_0) = 0. \tag{2}$$

The origin of (2) suggests a method of successive approximations. Let v_1, the new policy, be determined by the equation

$$T(u_0, v_1) = \max_v T(u_0, v), \tag{3}$$

and let u_1, the new return, be determined by the equation

$$T(u_1, v_1) = 0, \tag{4}$$

and so on.

In many important cases we can demonstrate that the convergence is monotonic, and quadratic. This will be the case below.

4. An Example

Consider the equation

$$u^2 - 2 = 0. \tag{1}$$

Let us employ the identity

$$u^2 = (v + (u - v))^2 = v^2 + 2v(u - v) + (u - v)^2. \tag{2}$$

From this it follows that

$$u^2 \geqslant v^2 + 2v(u - v) = 2uv - v^2, \tag{3}$$

for all v, and that

$$u^2 = \max_v [2uv - v^2], \tag{4}$$

with the maximum furnished by $v = u$.

Hence the nonlinear equation takes the form

$$\max_v [2uv - v^2 - 2] = 0. \tag{5}$$

From this it follows that

$$2uv - v^2 - 2 \leqslant 0, \tag{6}$$

for all v. If we are interested in the positive solution, we have

$$u \leqslant (v^2 + 2)/2v\,, \tag{7}$$

whence

$$u = \min_{v>0} [(v^2 + 2)/2v]\,. \tag{8}$$

If we employ (4) as a basis for a method of successive approximation, and proceed as above we are led to the celebrated recurrence relation

$$u_{n+1} = (u_n^{\,2} + 2)/2u_n\,. \tag{9}$$

This coincides with that obtained from the Newton–Raphson method, which assures us of quadratic convergence.

EXERCISES

1. Let $g(u)$ be convex. Show that $g(u) = \max_v [g(v) + (u-v)\, g'(v)]$.
2. What is the geometric interpretation of this analytic result?
3. What is the multidimensional version of the result of Exercise 1?
4. Consider the two relations

$$0 = f(u) = f(u_n) + f'(u_n)\,(u - u_n) + O[(u - u_n)^2]\,,$$
$$0 = f(u_n) + f'(u_n)\,(u_{n+1} - u_n)\,,$$

and thus deduce quadratic convergence to the root of $f(u) = 0$.

5. The Equation $u_{xx} + u_{yy} = u^2$

Let us now consider the numerical solution of the equation

$$u_{xx} + u_{yy} = u^2\,, \tag{1}$$

over a region R, where

$$u = f, \tag{2}$$

on Γ, the boundary of R. We will focus initially on this simple example since it will enable us to demonstrate conveniently the convergence properties of quasilinearization.

As we have noted, the most common method for establishing the existence of a solution of (1) and (2) is Picard iteration. We replace (1) by the sequence of linear problems

$$u_{xx}^{(i+1)} + u_{yy}^{(i+1)} = [u^{(i)}]^2, \tag{3}$$

where

$$u^{(i+1)} = f, \tag{4}$$

on Γ. Since $[u^{(i)}]^2$ is a forcing function for (3), we can examine the convergence of (3) via the Green's function representation of $u^{(i+1)}$,

$$u^{(i+1)}(x,y) = \iint_R k(x,y,a,b)\,[u^{(i)}(a,b)]^2\,da\,db. \tag{5}$$

Although we will not carry out the details, it can be shown that (3) does converge to (1) provided that $v^{(0)}$ is a sufficiently good initial approximation. However, the convergence is geometric. By this we mean that if we define a functional norm

$$\|g\| = \max_R |g(x,y)|, \tag{6}$$

then

$$\|u - u^{(i+1)}\| = O(\|u - u^{(i)}\|). \tag{7}$$

Thus Picard iteration does not in general lead to an efficient numerical method.

Let us now consider the quasilinear approximation to (1). We are led to the iterative method

$$u_{xx}^{(i+1)} + u_{yy}^{(i+1)} = 2u^{(i+1)}u^{(i)} - [u^{(i)}]^2, \tag{8}$$

with

$$u^{(i+1)} = f, \tag{9}$$

on Γ.

6. A Differential Inequality

We wish to relate solutions of the differential inequality

$$u_{xx} + u_{yy} + qu \geqslant 0, \tag{1}$$

subject to

$$u = f, \tag{2}$$

on the boundary, to the solution of the equation

$$v_{xx} + v_{yy} + qv = 0, \tag{3}$$

subject to

$$v = f, \tag{4}$$

on the boundary. We will assume q is such that the quadratic functional

$$H(u) = \int_R\int u_x^{\,2} + u_y^{\,2} - qu^2]\, dx\, dy\,, \tag{5}$$

is positive definite. A simple sufficient condition for this is $q \leqslant 0$.

For any function u which satisfies (1) and (2) we can write

$$u_{xx} + u_{yy} + qu - p = 0, \tag{6}$$

where

$$p \geqslant 0. \tag{7}$$

Then if we define w as

$$w = v - u, \tag{8}$$

w must satisfy

$$w_{xx} + w_{yy} + qw + p = 0, \tag{9}$$

where

$$w = 0, \tag{10}$$

on the boundary. Then since p is nonnegative and the associated Green's function is nonpositive (a result established in Section 6 of Chapter 1), we have

$$w \geqslant 0, \tag{11}$$

or the desired result

$$v(x,y) \geqslant u(x,y). \tag{12}$$

7. Monotonicity

We are now prepared to show that the sequence $u^{(i+1)}$ defined by

$$u_{xx}^{(i+1)} + u_{yy}^{(i+1)} = 2u^{(i+1)}u^{(i)} - [u^{(i)}]^2, \tag{1}$$

with

$$u^{(i+1)} = f, \tag{2}$$

on Γ, $v^{(0)} \equiv f$, converges monotonically, for $i \geqslant 1$, to the solution of

$$u_{xx} + u_{yy} = u^2, \tag{3}$$

with

$$u = f, \tag{4}$$

on Γ, provided that this solution exists. First we will prove

$$u \geqslant u^{(i)}, \qquad i = 1, 2, \ldots. \tag{5}$$

We have already indicated that at any point

$$2u^{(i)}u - [u^{(i)}]^2 \leqslant u^2. \tag{6}$$

Hence

$$u_{xx} + u_{yy} \geqslant 2uu^{(i)} - [u^{(i)}]^2. \tag{7}$$

Let us define a sequence of functions $w^{(i+1)}$ by

$$w^{(i+1)} = u - u^{(i+1)}. \tag{8}$$

Then by (1)–(4), $w^{(i+1)}$ satisfies

$$w_{xx}^{(i+1)} + w_{yy}^{(i+1)} - 2u^{(i)}w^{(i+1)} \geqslant 0, \tag{9}$$

with

$$w^{(i+1)} = 0, \tag{10}$$

on Γ. Using the result of Section 6, we have

$$w^{(i+1)} \geqslant 0 \tag{11}$$

and thus

$$u \geqslant u^{(i)}. \tag{12}$$

We will now establish a similar relation between $u^{(i+1)}$ and $u^{(i)}$. Taking $u = u^{(i+1)}$ in (3) and (6) we have

$$u_{xx}^{(i+1)} + u_{yy}^{(i+1)} \geqslant 2u^{(i+1)}u^{(i)} - [u^{(i+1)}]^2, \tag{13}$$

with

$$u^{(i+1)} = f, \tag{14}$$

on Γ. We now repeat the previous argument to find

$$u^{(i+1)} \geqslant u^{(i)}. \tag{15}$$

Combining (12) and (15) we have

$$u \geqslant u^{(i+1)} \geqslant u^{(i)} \geqslant \cdots \geqslant u^{(1)}, \tag{16}$$

or monotonic convergence.

8. Maximum Domain of Convergence

What is quite interesting about the foregoing is that we have convergence for any choice of the initial function provided that the equation has a solution. We thus have a maximum domain of convergence for the method of successive approximations we have employed. Note again that in this case quasilinearization yields the Newton–Raphson–Kantorovich approximation method, together with monotonicity.

9. Quadratic Convergence

As far as numerical computation is concerned, the most important property of quasilinearization is quadratic convergence. We will show that for the problem we are discussing

$$\| u - u^{(i+1)} \| \leqslant k \| u - u^{(i)} \|^2, \tag{1}$$

for some constant k. The iterative equation can be written as

$$u_{xx}^{(i+1)} + u_{yy}^{(i+1)} = [u^{(i)}]^2 + 2[u^{(i+1)} - u^{(i)}]\, u^{(i)}, \tag{2}$$

with

$$u^{(i+1)} = f, \tag{3}$$

on Γ. Since

$$u^2 = [u^{(i)}]^2 + 2u^{(i)}[u - u^{(i)}] + [u - u^{(i)}]^2, \tag{4}$$

the original equation

$$u_{xx} + u_{yy} = u^2, \tag{5}$$

can be written

$$u_{xx} + u_{yy} = [u^{(i)}]^2 + 2u^{(i)}[u - u^{(i)}] + [u - u^{(i)}]^2. \tag{6}$$

Subtracting (2) from (6) and defining $w^{(i)}$ by

$$w^{(i)} = u - u^{(i)}, \tag{7}$$

we have

$$w_{xx}^{(i+1)} + w_{yy}^{(i+1)} = 2u^{(i)} w^{(i+1)} + [w^{(i)}]^2, \tag{8}$$

with

$$w^{(i+1)} = 0 \tag{9}$$

on Γ. Using the Green's function, we have

$$w^{(i+1)}(x, y) = \iint_R k(x, y, x', y') \, [w^{(i)}(x', y')]^2 \, dx' \, dy'. \tag{10}$$

It is easy to show now that

$$w^{(i+1)}(x,y) \leqslant \max_R |w^i(x, y)| \iint_R |k(x, y, a, b)| \, da \, db. \tag{11}$$

The above result can now be written as, using (7),

$$\| u - u^{(i+1)} \| \leqslant k_1 \| u - u^{(i)} \|^2, \tag{12}$$

where k_1 is a constant.

10. Computational Aspects

The method of quasilinearization replaces the nonlinear elliptic equation

$$u_{xx} + u_{yy} = g(u), \tag{1}$$

by the sequence of linear problems

$$u_{xx}^{(i+1)} + u_{yy}^{(i+1)} = g(u^{(i)}) + (u^{(i+1)} - u^{(i)}) g_u(u^{(i)}), \tag{2}$$

where (2) is subject to the same boundary conditions as (1). The nonlinear parabolic equation

$$u_t = u_{xx} + u_{yy} + g(u) \tag{3}$$

is treated in exactly the same manner, i.e., replaced by the sequence

$$u_t^{(i+1)} = u_{xx}^{(i+1)} + u_{yy}^{(i+1)} + g(u^{(i)}) + (u^{(i+1)} - u^{(i)}) g_u(u^{(i)}). \tag{4}$$

For an arbitrary function $g(u)$, not necessarily convex or concave, we might not have the favorable convergence properties we have demonstrated for our sample problem. However, for well behaved functions $g(u)$, quasilinearization will converge if the initial guess, $u^{(0)}$, is sufficiently close to the solution. In most physical problems it is usually not a difficult problem to obtain a reasonably good initial guess. The problem usually is that of obtaining a reasonable degree of accuracy.

We must store the previous solution, $u^{(i)}$, in order to generate the succeeding solution $u^{(i+1)}$. Unless the grid size is made exceeding small, the amount of storage required for this is not significant.

11. Example

As a second numerical example let us consider the solution of

$$u_{xx} + u_{yy} = e^u , \tag{1}$$

on the rectangle $0 \leqslant x \leqslant \frac{1}{2}$, $0 \leqslant y \leqslant \frac{1}{4}$, subject to

$$u = 0 , \tag{2}$$

on the boundary. The quasilinearized version of (1) is

$$u_{xx}^{(i+1)} + u_{yy}^{(i+1)} = \exp(u^{(i)}) \, [u^{(i+1)} + (1 - u^{(i)})] . \tag{3}$$

The initial guess used was

$$u^{(0)} = 0 . \tag{4}$$

The method converged to four significant places as is shown in Table I for two typical points. The problem was solved again, this time with the boundary conditions

$$u = 10 , \tag{5}$$

and with the initial guess

$$u^{(0)} = 6 . \tag{6}$$

Four iterations were sufficient for convergence to four digits.

TABLE I

Iteration	$u(\frac{1}{4},\frac{1}{8})$	$u(\frac{1}{8},\frac{1}{8})$
0	0.0	0.0
1	−0.00707060	−0.0060304
2	−0.00707072	−0.00603050
3	−0.00707072	−0.00603050

This problem was previously solved using quasilinearization and successive overrelaxation. The optimal relaxation parameter was found to be different for (3) and (4) from the optimal relaxation factor for (3) and (5). Choosing a parameter suboptimally can cause the number of iterations to double. Furthermore, the optimal parameter was "optimal" only in the sense that it was kept constant for the entire problem. Certainly this choice is not optimal for any given iteration since the function $e^{u(i)}$ keeps changing from stage to stage. Thus, it is clear that iterative methods for solving linear elliptic equations are not particularly well suited for use with quasilinearization. However, the direct methods of dynamic programming and invariant imbedding seem to fit in perfectly with quasilinearization.

12. Identification Problems

Thus far we have been concerned with only one problem. Given a partial differential equation and a set of boundary conditions, we are required to obtain a numerical solution. Suppose, however, we are interested in some physical process in which we know that the variable of interest satisfies an equation of the form

$$u_{xx} + u_{yy} = g(u, a), \tag{1}$$

where a is an unknown parameter or vector of parameters. We are allowed to observe the physical process and to make measurements of u at various points. From our measurements we are required to determine the parameter a so that the solution of (1) agrees with the observed data.

In other variations of this inverse, or identification, problem we are required to identify initial or boundary conditions from observations. Inverse problems arise in such diverse settings as the study of drug distributions in the body, cardiology, and weather prediction.

13. The Least-Squares Criterion

We will concentrate on the following problem. We are given a physical process, assumed to be described by the equation

$$u_{xx} + u_{yy} + g(u, a) = 0, \tag{1}$$

where a,

$$a = (a_1, a_2, \ldots, a_k), \tag{2}$$

is an unknown vector parameter. We will assume that we know that

$$u = f, \tag{3}$$

on the boundary of the region of interest. Suppose we make a set of L observations of the process. Each observation consists of an observed value of u, u_l; the points of observation are (x_l, y_l).

As a measure of the deviation between the solution of (1) and (2) for a given a and the observations, we will use the least squares criterion

$$S = \sum_{l=1}^{L} [u(x_l, y_l) - u_l]^2 . \tag{4}$$

The identification problem now becomes one of minimizing S over all possible choices of a. Although there are many other ways of choosing a goodness of fit criterion, the favorable analytic properties of the least squares criterion make it a popular choice.

14. Newton–Raphson–Kantorovich Method

It is clear that if the boundary conditions and the function g are sufficiently smooth, the function S is a continuous function of a. To emphasize this point, we rewrite the problem as

$$\min_a S(a) = \min_a \sum_{l=1}^{L} [u(x_l, y_l, a) - u_l]^2 . \tag{1}$$

Differentiating the function in (1) with respect to a, we find that necessary conditions for a minimum are

$$\frac{\partial S}{\partial a_i} = S_{a_i} = 0, \qquad i = 1, 2, \ldots, k, \tag{2}$$

or using the gradient

$$\nabla S(a) = 0 . \tag{3}$$

Our method will be to solve (2) [or (3)] by the Newton–Raphson–Kantorovich method.

Let us define the vector function $f(a)$ by

$$f(a) = \nabla S(a), \tag{4}$$

where we have

$$f(a) = [f_i(a)], \qquad i = 1, 2, \ldots, k . \tag{5}$$

We start by writing the Taylor series expansion of $f(a)$ about a vector b,

$$\begin{aligned} f_1(a) &= f_1(b) + (a_1 - b_1)\,[\partial f_1(b)/\partial a_1] + (a_2 - b_2)\,[\partial f_1(b)/\partial a_2] \\ &\quad + \cdots + (a_k - b_k)\,[\partial_k f(b)/\partial a_k] + \cdots, \\ f_2(a) &= f_2(b) + (a_1 - b_1)\,[\partial f_2(b)/\partial a_1] + \cdots, \\ f_k(a) &= f_k(b) + (a_1 - b_1)\,[\partial f_1(b)/\partial a_1] + \cdots. \end{aligned} \tag{6}$$

Keeping only the linear terms, we have the approximation

$$f(a) \cong f(b) + J(b)\,(a - b) \tag{7}$$

where J is the Jacobian matrix

$$J(b) = [\partial f_i(b)/\partial a_j]. \tag{8}$$

Since we are interested in a root of $f(a)$,

$$f(a) = 0, \tag{9}$$

our approximation, (7), yields

$$a = b - J(b)^{-1} f(b). \tag{10}$$

Since this value of a will be an approximation to the actual root, we use (10) iteratively,

$$a^{(i+1)} = a^{(i)} - J^{-1}(a^{(i)})\,f(a^{(i)}). \tag{11}$$

We expect quadratic convergence, that is

$$\| a - a^{(i+1)} \| = O(\| a - a^{(i)} \|^2). \tag{12}$$

Since $f(a)$ is given as

$$f(a) = \nabla S(a) \tag{13}$$

the Jacobian becomes the Hessian matrix

$$J(a) = (S_{a_i a_j}). \tag{14}$$

Thus each iteration of (11) requires the evaluation of $S_{a_i a_j}$ and S_{a_i} for $i = 1, \ldots, k$. We will now discuss how these functions are evaluated.

15. The Sensitivity Equations

Since we must evaluate $S_{a_i a_j}$ we will write out these functions in detail. Starting with

$$S = \sum_{l=1}^{L} [u(x_l, y_l) - u_l]^2 \tag{1}$$

we differentiate to find

$$S'_{a_i} = 2\sum_{l=1}^{L} [u(x_l, y_l) - u_l]\, u_{a_i}(x_l, y_l) \tag{2}$$

and

$$S_{a_i a_j} = 2\sum_{l=1}^{L} [u_{a_i}(x_l, y_l)\, u_{a_j}(x_l, y_l) - (u(x_l, y_l) - u_l)\, u_{a_i a_j}(x_l, y_l)]\,. \tag{3}$$

Thus, we have introduced the sensitivity functions u_{a_j} and $u_{a_i a_j}$ which must be evaluated at each observation point. We will now show that the sensitivity functions satisfy elliptic equations of a special nature.

We start with the original equation

$$u_{xx} + u_{yy} + g(u, a) = 0\,. \tag{4}$$

Differentiation with respect to a_i yields

$$(u_{a_i})_{xx} + (u_{a_i})_{yy} + g_u(u, a)\, u_{a_i} + g_{a_i}(u, a_i) = 0\,, \tag{5}$$

Differentiating again we have

$$(u_{a_i a_j})_{xx} + (u_{a_i a_j})_{yy} + g_u(u, a)\, u_{a_i a_j} + g_{uu}(u, a)\, u_{a_i} u_{a_j} + g_{ua_j}(u, a)\, u_{a_i} + g_{a_i u}(u, a)\, u_{a_j} + g_{a_i a_j}(u, a) = 0\,. \tag{6}$$

Since the boundary conditions on u are fixed, we have

$$u_{a_i} = 0\,, \tag{7}$$

and

$$u_{a_i a_j} = 0\,, \tag{8}$$

on the boundaries. We note that these $L^2 + L$ sensitivity equations are *linear* and are all of the form

$$v_{xx} + v_{yy} + g_u(u, a)\, v + q = 0\,, \tag{9}$$

where only the function q changes in different equations. We will exploit this similarity in a moment.

16. Quasilinearization

In general the original equation

$$u_{xx} + u_{yy} + g(u, a) = 0 \tag{1}$$

is nonlinear. Thus, we employ quasilinearization to obtain a numerical solution. We replace (1) by the sequence of linear equations

$$u_{xx}^{(i+1)} + u_{yy}^{(i+1)} + g_u(u^{(i)}, a)\,[u^{(i+1)} - u^{(i)}] + g(u^{(i)}, a) = 0\,. \tag{2}$$

This equation, as are the sensitivity equations, is also of the form

$$v_{xx} + v_{yy} + g_u(u, v)\,v + p = 0\,. \tag{3}$$

Thus, as the sequence defined by (2) converges, (2) and all the sensitivity equations differ only in their forcing functions and boundary conditions. This is a tremendous computational advantage.

In Chapters 5 and 6 we saw that the numerical solution of an equation of the form (3) involved the solution of matrix and vector recurrence relations of the form

$$\begin{aligned} A_{R-1} &= [2I+Q+F_R-A_R]^{-1}\,, \\ b_{R-1} &= A_{R-1}(b_R+r_R+p_R)\,, \end{aligned} \tag{4}$$

starting with the initial conditions

$$A_N = 0\,, \qquad b_N = v_N\,, \tag{5}$$

where F_R is derived from g_u, r_R is derived from the boundary conditions and p_R is derived from p. Thus, it is clear that since the A_{R-1} are independent of both the forcing function, p, and the boundary conditions, the matrices A_R are the same for the last quasilinear iteration and for all the L^2+L sensitivity equations. The solution of the matrix recurrence equation

$$A_{R-1} = [2I + Q + F_R - A_R]^{-1}\,, \tag{6}$$

is thus computed once, stored, and used to generate solutions of all the sensitivity equations. Since for each sensitivity equation we now need only to compute the b_R by (4) and the solution via

$$v_{i+1} = A_i v_i + b_i\,, \tag{7}$$

we have eliminated most of the computation required.

17. Example

Let us consider an example of the general problem just considered. Given the nonlinear equation

$$u_{xx} + u_{yy} + ae^{bu} = 0\,, \tag{1}$$

where

$$u = g\,, \tag{2}$$

on the boundary. We would like to find a choice of a and b such that

$$S(a,b) = \sum_{l=1}^{L} [u(x_l, y_l) - u_l]^2\,, \tag{3}$$

is minimum. First we replace (1) by the sequence of linear problems, for any fixed a and b,

$$u_{xx}^{(n+1)} + u_{yy}^{(n+1)} + a\exp(bu^{(n)}) + ab(u^{(n+1)} - u^{(n)})\exp(bu^{(n)}) = 0\,, \tag{4}$$

subject to

$$u^{(n+1)} = g\,, \tag{5}$$

on the boundary. The optimal a and b are given by the solution of

$$S_a = 0\,, \qquad S_b = 0\,, \tag{6}$$

Then using the Newton–Raphson–Kantorovich method we generate the sequences $a^{(m)}$ and $b^{(m)}$ by means of

$$\begin{bmatrix} a^{(m+1)} \\ b^{(m+1)} \end{bmatrix} = \begin{bmatrix} a^{(m)} \\ b^{(m)} \end{bmatrix} - J(a^{(m)}, b^{(m)})^{-1} \begin{bmatrix} S_a(a^{(m)}, b^{(m)}) \\ S_b(a^{(m)}, b^{(m)}) \end{bmatrix}, \tag{7}$$

where

$$J(a,b) = \begin{bmatrix} S_{aa}(a,b) & S_{ab}(a,b) \\ S_{ba}(a,b) & S_{bb}(a,b) \end{bmatrix}. \tag{8}$$

Differentiation of (3) yields

$$\begin{aligned} S_a &= 2\sum_{l=1}^{L} [u(x_l, y_l) - u_l]\, u_a(x_l, y_l)\,, \\ S_b &= 2\sum_{l=1}^{L} [u(x_l, y_l) - u_l]\, u_b(x_l, y_l)\,, \end{aligned} \tag{9}$$

and

$$\begin{aligned} S_{aa} &= 2\sum_{l=1}^{L} [u(x_l, y_l) - u_l]\, u_{aa}(x_l, y_l) + [u_a(x_l, y_l)]^2\,, \\ S_{bb} &= 2\sum_{l=1}^{L} [u(x_l, y_l) - u_l]\, u_{bb}(x_l, y_l) + [u_b(x_l, y_l)]^2\,, \\ S_{ab} &= S_{ba} = 2\sum_{l=1}^{L} [u(x_l, y_l) - u_l]\, u_{ab}(x_l, y_l) + u_a(x_l, y_l)\, u_b(x_l, y_l)\,. \end{aligned} \tag{10}$$

The necessary sensitivity equations are found by differentiation of (1) to be

$$
\begin{aligned}
(u_a)_{xx} + (u_a)_{yy} + e^{bu} + abe^{bu}u_a &= 0\,, \\
(u_b)_{xx} + (u_b)_{yy} + aue^{bu} + abe^{bu}u_b &= 0\,,
\end{aligned}
\tag{11}
$$

and

$$
\begin{aligned}
(u_{aa})_{xx} + (u_{aa})_{yy} + 2be^{bu}u_a + abe^{bu}u_{aa} + ab^2e^{bu}u_a{}^2 &= 0\,, \\
(u_{bb})_{xx} + (u_{bb})_{yy} + 2ae^{bu}u_b + 2abue^{u}u_b \qquad & \\
+ au^2e^{bu} + abe^{bu}u_{bb} + ab^2e^{bu}u_b{}^2 &= 0\,, \\
(u_{ab})_{xx} + (u_{ab})_{yy} = ue^{bu} + be^{bu}u_b + ae^{bu}u_a \qquad & \\
+ abue^{bu}u_a + abe^{bu}u_{ab} + ab^2e^{bu}u_au_b &= 0\,, \\
u_{ba} &= u_{ab}\,,
\end{aligned}
\tag{12}
$$

with

$$
u_a = u_b = u_{aa} = u_{bb} = u_{ab} = 0 \tag{13}
$$

on the boundary. Note that all the equations of (11) and (12) are of the form

$$
v_{xx} + v_{yy} + abe^{bu}v + p = 0\,. \tag{14}
$$

The procedure is as follows. For a given $a = a^{(m)}$ and $b = b^{(m)}$ we solve (1) via the quasilinear equation (4). As the solution, $u^{(n+1)}(x, y, a^{(m)}, b^{(m)})$ converges, we save the matrices A_R, from the final iteration of (4). The sensitivity equations of (11) and (12) are solved using these matrices and the last value of $u^{(n+1)}$ as u to generate the necessary functions. Then (9) and (10) are evaluated and (7) is solved to give

$$
\begin{aligned}
a^{(m+1)} &= a^{(m)} \\
&\quad - \frac{S_{bb}(a^{(m)}, b^{(m)})\, S_a(a^{(m)}, b^{(m)}) - S_{ab}(a^{(m)}, b^{(m)})\, S_b(a^{(m)}, b^{(m)})}{S_{aa}(a^{(m)}, b^{(m)})\, S_{bb}(a^{(m)}, b^{(m)}) - [S_{ab}(a^{(m)}, b^{(m)})]^2}\,, \\
b^{(m+1)} &= b^{(m)} \\
&\quad - \frac{S_{ab}(a^{(m)}, b^{(m)})\, S_a(a^{(m)}, b^{(m)}) - S_{bb}(a^{(m)}, b^{(m)})\, S(a^{(m)}, b^{(m)})}{S_{aa}(a^{(m)}, b^{(m)})\, S_{bb}(a^{(m)}, b^{(m)}) - [S_{ab}(a^{(m)}, b^{(m)})]^2}\,.
\end{aligned}
\tag{15}
$$

The procedure is repeated until the sequences for $a^{(m)}$ and $b^{(m)}$ converge. At each iteration we use the last solution as an initial guess, i.e.,

$$
u^{(0)}(x, y, a^{(m+1)}, b^{(m+1)}) = u^{(\bar{n})}(x, y, a^{(m)}, b^{(m)})\,, \tag{16}
$$

where $\bar{n}$ refers to the final iteration of the previous quasi-linear iteration.

This problem was solved numerically on a unit square using $(15^2) = 225$ interior points. The results are shown in Table II. The entire computation took 8 seconds of computer time.

TABLE II

Iteration	$a(k)$	$b(k)$
0	1.5000	1.5000
1	1.6819	2.2737
2	1.6805	2.2686
3	1.6806	2.2687
4	1.6806	2.2687

MISCELLANEOUS EXERCISES

1. Introduce the stationary operation, S, defined in the following fashion, $S_u(f(u)) = f(v)$, where v is the assumed unique solution of $f'(u) = 0$. Show that if $f'(u)$ is never zero, then

$$f(u) = S_v[f(v) + (u - v)f'(v)].$$

When is the operation one of maximization and when of minimization?

2. Show that if, in addition, $f'(u)$ is never zero, then we can write

$$r = S_v[v - f(v)/f'(v)]$$

as a representation for r, the assumed unique solution of $f(u) = 0$.

BIBLIOGRAPHY AND COMMENT

Section 1. See

R. Bellman and R. Kalaba, *Quasilinearization and Nonlinear Boundary-Value Problems*, Amer. Elsevier, New York, 1965.

Section 11. See

R. Bellman, M. Juncosa, and R. Kalaba, "Some Numerical Experiments Using Newton's Method for Nonlinear Parabolic and Elliptic Boundary-Value Problems," *Comm. ACM*, Vol. 4, 1961, pp. 187–191.

Section 6. See

E. F. Beckenbach and R. Bellman, *Inequalities*, Springer-Verlag, Berlin and New York, 1961.

Section 12. See

J. Casti, D. Detchmendy, H. Kagiwada, and R. Kalaba, "Estimating the Parameters of an Inhomogeneous Medium by Probing with Rays," in *Computational Approaches in Applied Mechanics*, Amer. Soc. of Mech. Eng., New York, 1969, pp. 200–209.

E. Angel, "Inverse Boundary-Value Problems for Elliptic Equations," *J. Math. Anal. Appl.*, Vol. 30, 1970, pp. 86–98.

Appendix

Computer Programs

This appendix contains four sample computer programs and the output from these programs. Although constant coefficient equations are solved, we make no use of the simplifications of Chapter 8. Thus these programs can be easily generalized to include a large class of equations.

The following subroutines are necessary:

$MV(A, N, B, C)$ computes product of square matrix A and N-dimensional vector B and puts result in C.

$MM(A, N, B, C)$ computes matrix product AB and puts result in C. Matrices are square and of order N.

$MINV(A, N, D, L, M)$—IBM scientific subpackage Gauss–Jordan—matrix A of order N is inverted in place. Determinant is put in D. L and M are N-dimensional work vectors.

Program 1. Dynamic Programming

Dynamic programming is used to solve the potential equation

$$u_{xx} + u_{yy} = 0$$

over a square region. The matrices $I - A_R$ and vectors b_R, as described in Chapter 5, are generated and stored on a high speed disc. The significant equations are

$$A_R = I - [I + Q - A_{R+1}]^{-1}, \qquad A_N = I,$$

$$b_R = [I - A_R](b_{R+1} + r_R), \qquad b_N = u_N,$$

and

$$u_R = [I - A_R]u_{R-1} + b_R.$$

```
      DIMENSION A(15,15),B(15),LL(15),LM(15),BB(15),U(15),
     1UL(17),UR(17),UT(15),UB(15)
      DEFINE FILE 1(240,360,L,ID)
C
C  INITIALIZATION
C
      ID=1
      READ(5,100) N,M
  100 FORMAT(2I2)
      READ(5,101) UL
      READ(5,101) UR
      READ(5,101) UT
      READ(5,101) UB
  101 FORMAT(9F6.3,/,8F6.3)
      WRITE(6,103) N,M
  103 FORMAT('1 SOLUTION OF POTENTIAL EQUATION BY DYNAMIC PROGRAMMING',/
     1/,I10,' POINTS IN X DIRECTION',/,I10,' POINTS IN Y DIRECTION',//)
      DO 1 I=1,M
      DO 1 J=1,M
    1 A(I,J)=0.
      DO 2 I=1,M
      A(I,I)=1.
    2 E(I)=UR(I+1)
      DO 3 K=1,N
      NN=N-K+1
C
C  COMPUTE I-A(I-1)
C
      DO 4 I=1,M
    4 A(I,I)=A(I,I)+3.
      DO 5 I=2,M
      A(I,I-1)=A(I,I-1)-1.
    5 A(I-1,I)=A(I-1,I)-1.
      CALL MINV(A,M,D,LL,LM)
      IF(D.EQ.0.) STOP
C
C  COMPUTE B(I-1)
C
      DO 6 I=1,M
    6 BB(I)=B(I)
      BB(1)=BB(1)+UB(NN)
      BB(M)=BB(M)+UT(NN)
      CALL MV(A,M,BB,B)
      IF(K.EQ.N) GO TO 7
```

```
C
C   STORE I-A(I-1) AND B(I-1) ON DISC
C
         DO 8 I=1,M
       8 WRITE(1'ID) (A(I,J),J=1,M)
         WRITE(1'ID) (B(I),I=1,M)
C
C   COMPUTE A(I-1)
C
         DO 9 I=1,M
         DO 9 J=1,M
       9 A(I,J)=-A(I,J)
         DO 3 I=1,M
       3 A(I,I)=A(I,I)+1.
       7 ID=ID+M+1
         WRITE(6,102) UL
     102 FORMAT(1X,9F12.5,/,10X,8F12.5)
         DO 10 I=1,M
      10 BB(I)=UL(I+1)
         CALL MV(A,M,BB,U)
         DO 11 I=1,M
      11 U(I)=U(I)+B(I)
         WRITE(6,102) UB(1),U,UT(1)
         DO 12 K=2,N
C
C   RECALL I-A(I) AND B(I)
C
         ID=ID-2*(M+1)
         DO 13 I=1,M
      13 READ(1'ID) (A(I,J),J=1,M)
         READ(1'ID) (B(J),J=1,M)
C
C   COMPUTE U(I)
C
         CALL MV(A,M,U,BB)
         DO 14 I=1,M
      14 U(I)=BB(I)+B(I)
      12 WRITE(6,102) UB(K),U,UT(K)
         WRITE(6,110) UR
     110 FORMAT(1X,9F12.5,/,10X,8F12.5,/,'1')
         STOP
         END
```

```
SOLUTION OF POTENTIAL EQUATION BY DYNAMIC PROGRAMMING
  15 POINTS IN X DIRECTION
  15 POINTS IN Y DIRECTION

0.0        0.0       0.0       0.0       0.0       0.0       0.0       0.0       0.0
         0.0       0.0       0.0       0.0       0.0       0.0       0.0       0.0
0.0        0.00429   0.00876   0.01359   0.01898   0.02517   0.03246   0.04124   0.05205
         0.06570   0.08346   0.10757   0.14231   0.19691   0.29386   0.49570   1.00000
0.0        0.00842   0.01717   0.02662   0.03715   0.04924   0.06343   0.08045   0.10127
         0.12728   0.16057   0.20450   0.26476   0.35149   0.48283   0.68895   1.00000
0.0        0.01220   0.02488   0.03856   0.05378   0.07120   0.09157   0.11586   0.14531
         0.18158   0.22704   0.28509   0.36077   0.46144   0.59701   0.77729   1.00000
0.0        0.01550   0.03160   0.04894   0.06822   0.09020   0.11580   0.14612   0.18251
         0.22670   0.28092   0.34804   0.43177   0.53649   0.66647   0.82321   1.00000
0.0        0.01819   0.03708   0.05740   0.07995   0.10558   0.13531   0.17029   0.21191
         0.26180   0.32188   0.39440   0.48179   0.58629   0.70917   0.84907   1.00000
0.0        0.02018   0.04113   0.06365   0.08858   0.11687   0.14956   0.18784   0.23305
         0.28670   0.35042   0.42590   0.51468   0.61772   0.73486   0.86389   1.00000
0.0        0.02140   0.04361   0.06747   0.09386   0.12376   0.15823   0.19846   0.24576
         0.30152   0.36720   0.44412   0.53329   0.63507   0.74864   0.87165   1.00000
0.0        0.02181   0.04445   0.06876   0.09564   0.12607   0.16114   0.20201   0.24999
         0.30644   0.37272   0.45007   0.53932   0.64061   0.75299   0.87407   1.00000
0.0        0.02140   0.04361   0.06747   0.09386   0.12376   0.15823   0.19846   0.24576
         0.30152   0.36720   0.44412   0.53330   0.63507   0.74864   0.87165   1.00000
0.0        0.02018   0.04113   0.06365   0.08858   0.11687   0.14956   0.18784   0.23305
         0.28670   0.35042   0.42591   0.51468   0.61773   0.73486   0.86389   1.00000
0.0        0.01819   0.03708   0.05740   0.07995   0.10558   0.13531   0.17030   0.21192
         0.26180   0.32189   0.39441   0.48179   0.58630   0.70918   0.84907   1.00000
0.0        0.01550   0.03160   0.04895   0.06822   0.09020   0.11580   0.14612   0.18251
         0.22670   0.28092   0.34805   0.43178   0.53650   0.66647   0.82321   1.00000
0.0        0.01220   0.02488   0.03856   0.05378   0.07120   0.09157   0.11586   0.14531
         0.18158   0.22704   0.28509   0.36077   0.46144   0.59701   0.77729   1.00000
0.0        0.00842   0.01717   0.02662   0.03715   0.04924   0.06343   0.08045   0.10127
         0.12728   0.16057   0.20450   0.26477   0.35149   0.48283   0.68896   1.00000
0.0        0.00429   0.00876   0.01359   0.01898   0.02517   0.03246   0.04124   0.05205
         0.06570   0.08346   0.10757   0.14231   0.19692   0.29386   0.49570   1.00000
0.0        0.0       0.0       0.0       0.0       0.0       0.0       0.0       0.0
         0.0       0.0       0.0       0.0       0.0       0.0       0.0       0.0
```

Program 2. Riccati Transformation

Once again the potential equation is solved. In this case only rapid access storage is used. Thus, as is to be expected, this program executes faster than the previous one. The significant equations are

$$A_R = [2I + Q - A_{R+1}]^{-1}, \qquad A_{N-1} = 0,$$

$$b_R = A_R(b_R + r_R), \qquad b_{N-1} = u_N,$$

and

$$u_{R+1} = A_R u_R + b_R.$$

```
      DIMENSION A(15,15,16),B(15,16),U(15),BB(15),UL(17),UR(17),
     1UT(15),UB(15),LL(15),LM(15)
C
C  INITIALIZATION
C
      READ(5,100) M,N
  100 FORMAT(2I2)
      READ(5,101) UL
      READ(5,101) UR
      READ(5,101) UT
      READ(5,101) UB
  101 FORMAT(9F6.3,/,8F6.3)
      WRITE(6,103) N,M
  103 FORMAT('1 SOLUTION OF POTENTIAL EQUATION BY THE RICCATI TRANSFORMA
     1TION',//,I10,' POINTS IN X DIRECTION',/,I10,' POINTS IN Y DIRECTIO
     2N',//)
      DO 1 I=1,M
      B(I,M+1)=UR(I+1)
      DO 1 J=1,M
    1 A(I,J,N+1)=0.
      DO 2 KK=1,N
      K=N-KK+1
C
C  COMPUTE A(I+1) FROM A(I)
C
      DO 3 I=1,M
      DO 3 J=1,M
    3 A(I,J,K)=-A(I,J,K+1)
      DO 4 I=1,M
    4 A(I,I,K)=A(I,I,K)+4.
      DO 5 I=2,M
      A(I,I-1,K)=A(I,I-1,K)-1.
    5 A(I-1,I,K)=A(I-1,I,K)-1.
      CALL MINV(A(1,1,K),M,D,LL,LM)
      IF(D.EQ.0.) STOP
```

```
C
C   COMPUTE B(I+1) FROM A(I+1) AND B(I)
C
          DO 6 I=1,M
        6 BB(I)=B(I,K+1)
          BB(1)=BB(1)+UB(K)
          BB(M)=BB(M)+UT(K)
        2 CALL MV(A(1,1,K),M,BB,B(1,K))
          WRITE(6,102) UL
      102 FORMAT(1X,9F12.5,/,10X,8F12.5)
          DO 7 I=1,M
        7 U(I)=UL(I+1)
          DO 8 K=1,N
C
C   COMPUTE U(I+1) FROM A(I),B(I) AND U(I)
C
          CALL MV(A(1,1,K),M,U,BB)
          DO 9 I=1,M
        9 U(I)=BB(I)+B(I,K)
        8 WRITE(6,102) UB(K),U,UT(K)
          WRITE(6,110) UR
      110 FORMAT(1X,9F12.5,/,10X,8F12.5,/,'1')
          STOP
          END
```

SOLUTION OF POTENTIAL EQUATION BY THE RICCATI TRANSFORMATION
15 POINTS IN X DIRECTICN
15 POINTS IN Y DIRECTICN

1.00000	1.00000	1.00000	1.00000	1.00000	1.00000	1.00000	1.00000	1.00000
	1.00000	1.00000	1.00000	1.00000	1.00000	1.00000	1.00000	1.00000
0.0	0.49570	0.68896	0.77729	0.82321	0.84907	0.86389	0.87165	0.87407
	0.87165	0.86389	0.84907	0.82321	0.77729	0.68896	0.49570	0.0
0.0	0.29386	0.48283	0.59700	0.66647	0.70917	0.73485	0.74864	0.75299
	0.74864	0.73485	0.70917	0.66647	0.59701	0.48283	0.29386	0.0
0.0	0.19691	0.35148	0.46143	0.53649	0.58629	0.61772	0.63506	0.64060
	0.63506	0.61772	0.58629	0.53649	0.46144	0.35149	0.19691	0.0
0.0	0.14231	0.26476	0.36076	0.43177	0.48176	0.51467	0.53329	0.53931
	0.53329	0.51467	0.48178	0.43177	0.36076	0.26476	0.14231	0.0
0.0	0.10757	0.20449	0.28508	0.34804	0.39440	0.42590	0.44411	0.45006
	0.44411	0.42590	0.39440	0.34804	0.28508	0.20450	0.10757	0.0
0.0	0.08346	0.16057	0.22704	0.28092	0.32188	0.35042	0.36719	0.37272
	0.36719	0.35042	0.32188	0.28092	0.22704	0.16057	0.08346	0.0
0.0	0.06570	0.12728	0.18158	0.22670	0.26180	0.28670	0.30152	0.30644
	0.30152	0.28670	0.26180	0.22670	0.18158	0.12728	0.06570	0.0
0.0	0.05205	0.10127	0.14530	0.18251	0.21191	0.23305	0.24575	0.24999
	0.24575	0.23305	0.21191	0.18251	0.14531	0.10127	0.05205	0.0
0.0	0.04124	0.08045	0.11586	0.14612	0.17029	0.18784	0.19846	0.20201
	0.19846	0.18784	0.17030	0.14612	0.11586	0.08045	0.04124	0.0
0.0	0.03246	0.06343	0.09157	0.11580	0.13531	0.14956	0.15823	0.16113
	0.15823	0.14956	0.13531	0.11580	0.09157	0.06343	0.03246	0.0
0.0	0.02517	0.04924	0.07120	0.09020	0.10558	0.11687	0.12376	0.12607
	0.12376	0.11687	0.10558	0.09020	0.07120	0.04924	0.02517	0.0
0.0	0.01898	0.03715	0.05378	0.06822	0.07995	0.08858	0.09386	0.09564
	0.09386	0.08858	0.07995	0.06822	0.05378	0.03715	0.01898	0.0
0.0	0.01359	0.02662	0.03856	0.04894	0.05740	0.06365	0.06747	0.06876
	0.06747	0.06365	0.05740	0.04894	0.03856	0.02662	0.01359	0.0
0.0	0.00876	0.01717	0.02488	0.03160	0.03708	0.04113	0.04361	0.04445
	0.04361	0.04113	0.03708	0.03160	0.02488	0.01717	0.00876	0.0
0.0	0.00429	0.00842	0.01220	0.01550	0.01819	0.02018	0.02140	0.02181
	0.02140	0.02018	0.01819	0.01550	0.01220	0.00842	0.00429	0.0
0.0	0.0	0.0	0.0	0.0	0.0	0.0	0.0	0.0
	0.0	0.0	0.0	0.0	0.0	0.0	0.0	0.0

Program 3. Invariant Imbedding

The potential equation is again solved numerically. In this example the solution in the middle of the region is obtained using the same grid spacing employed in the previous two examples. Note that the one sweep nature of the computation has eliminated the storage requirement. Thus we solve

$$R_i = [2I + Q - R_{i-1}]^{-1}, \qquad R_0 = 0,$$

$$S_i = R_i[s_{i-1} + r_i], \qquad s_0 = d,$$

and at the desired k we adjoin

$$U_{i+1} = U_i R_i, \qquad U_k = R_k,$$

$$p_{i+1} = p_i + U_i s_i, \qquad p_k = 0.$$

Finally we compute the product

$$u_k = U_N c + p_N.$$

```
0001          DIMENSION R(15,15),U(15,15),V(15,15),S(15),W(15),P(15),
             1UL(15),UR(15),UT(15),UB(15),IW(15),JW(15)
        C
        C   INITIALIZATICN
        C
0002          READ(5,100) M,N,NN
0003      100 FORMAT(3I2)
0004          READ(5,101) UL
0005          READ(5,101) UR
0006          READ(5,101) UT
0007          READ(5,101) UB
0008      101 FORMAT(15F4.2)
0009          WRITE(6,102) N,M
0010      102 FORMAT('1 SOLUTION OF POTENTIAL EQUATION BY INVARIANT IMBEDDING',/
             1,I10,' POINTS IN X DIRECTION',/,I10,' POINTS IN Y DIRECTION')
0011          WRITE(6,105) UT,UB,UL,UR
0012      105 FORMAT(/,' BOUNDARY VALUES UT,UB,UL,UR',(/,1X,8F12.5,/,5X,7F12.5))
0013          DO 1 I=1,M
0014          P(I)=0.
0015          S(I)=UL(I)
0016          DO 1 J=1,M
0017          U(I,J)=0.
0018        1 R(I,J)=0.
0019          DO 9 I=1,M
0020        9 U(I,I)=1.
0021          DO 2 K=1,N
        C
        C   COMPUTE R(I+1) FROM R(I)
        C
0022          DO 3 I=1,M
0023          DO 3 J=1,M
0024        3 R(I,J)=-R(I,J)
0025          DO 4 I=1,M
0026        4 R(I,I)=R(I,I)+4.
0027          DO 5 I=2,M
0028          R(I,I-1)=R(I,I-1)-1.
0029        5 R(I-1,I)=R(I-1,I)-1.
0030          CALL MINV(R,M,DET,IW,JW)
0031          IF(DET) 13,6,13
```

```
C
C  COMPUTE S(I+1) FROM R(I+1) AND S(I)
C
   13 DO 7 I=1,M
    7 UL(I)=S(I)
      UL(1)=UL(1)+UB(K)
      UL(M)=UL(M)+UT(K)
      CALL MV(R,M,UL,S)
      IF(K-NN) 2,14,14
C
C  ADJOIN EQUATIONS FOR U(I) AND P(I)
C
   14 CALL MV(U,M,S,UL)
      DO 11 I=1,M
   11 P(I)=P(I)+UL(I)
      CALL MM(R,M,U,V)
      DO 10 I=1,M
      DO 10 J=1,M
   10 U(I,J)=V(I,J)
    2 CONTINUE
C
C  COMPUTE SOLUTION VECTOR
C
      CALL MV(U,M,UR,W)
      DO 12 I=1,M
   12 W(I)=P(I)+W(I)
      WRITE(6,110) NN
  110 FORMAT(//,' SOLUTION AT X=',I3,//)
      WRITE(6,104) W
  104 FORMAT(//,1X,8F12.5,/,5X,7F12.5,/,'1')
    6 STOP
      END
```

```
SOLUTION OF POTENTIAL EQUATION BY INVARIANT IMBEDDING
      15 POINTS IN X DIRECTION
      15 POINTS IN Y DIRECTION

BOUNDARY VALUES UT,UB,UL,UR
     1.00000      1.00000      1.00000      1.00000      1.00000      1.00000      1.00000      1.00000
          1.00000      1.00000      1.00000      1.00000      1.00000      1.00000      1.00000

     0.0          0.0          0.0          0.0          0.0          0.0          0.0          0.0
          0.0          0.0          0.0          0.0          0.0          0.0          0.0

     0.0          0.0          0.0          0.0          0.0          0.0          0.0          0.0
          0.0          0.0          0.0          0.0          0.0          0.0          0.0

     0.0          0.0          0.0          0.0          0.0          0.0          0.0          0.0
          0.0          0.0          0.0          0.0          0.0          0.0          0.0

SOLUTION AT X=  8


     0.02181      0.04445      0.06876      0.09564      0.12607      0.16114      0.20201      0.24999
          0.30644      0.37272      0.45007      0.53932      0.64061      0.75299      0.87407
```

Program 4. Quasilinearization

The Riccati transformation is combined with quasilinearization to solve the nonlinear equation

$$u_{xx} + u_{yy} = u^2 .$$

Thus the program solves the sequence of linear equations

$$u_{xx}^{(i+1)} + u_{yy}^{(i+1)} - 2u^{(i)}[u^{(i+1)}] = -[u^{(i)}]^2 .$$

The initial guess

$$u^{(0)} \equiv 0$$

is used. Only the first three iterations are shown since succeeding iterations show no changes in the first five significant digits.

```
C
C   QUASILINEARIZATION AND RICCATI TRANSFORMATION
C
      DIMENSION A(15,15,16),B(15,16),U(15,15),BB(15),UL(17),UR(17),
     1UT(15),UB(15),LL(15),LM(15),UU(15)
      G(X)=X*X
      GU(X)=2.*X
      READ(5,100) M,N,H
  100 FORMAT(2I2,F10.5)
      READ(5,101) UL
      READ(5,101) UR
      READ(5,101) UT
      READ(5,101) UB
  101 FORMAT(9F6.3,/,8F6.3)
      DO 1 I=1,N
    1 READ(5,101) (U(I,J),J=1,M)
      ITER=0
      WRITE(6,102) ITER
  102 FORMAT(1H1,' ITERATION ',I2,///)
      WRITE(6,103) UL
  103 FORMAT(1X,9F12.5,/,10X,8F12.5)
      DO 2 I=1,N
    2 WRITE(6,103) UB(I),(U(I,J),J=1,M),UT(I)
      WRITE(6,103) UR
      DO 3 I=1,M
      B(I,N+1)=UR(I+1)
      DO 3 J=1,M
    3 A(I,J,N+1)=0.
   13 ITER=ITER+1
      DO 4 KK=1,N
      K=N-KK+1
      DO 5 I=1,M
      DO 5 J=1,M
    5 A(I,J,K)=-A(I,J,K+1)
```

```
      DO 6 I=1,M
    6 A(I,I,K)=A(I,I,K)+4.+H*H*GU(U(K,I))
      DO 7 I=2,M
      A(I,I-1,K)=A(I,I-1,K)-1.
    7 A(I-1,I,K)=A(I-1,I,K)-1.
      CALL MINV(A(1,1,K),M,D,LL,LM)
      IF(D.EQ.0.) STOP
      DO 8 I=1,M
    8 BB(I)=B(I,K+1)+H*H*(U(K,I)*GU(U(K,I))-G(U(K,I)))
      BB(1)=BB(1)+UB(K)
      BB(M)=BB(M)+UT(K)
    4 CALL MV(A(1,1,K),M,BB,B(1,K))
      WRITE(6,102) ITER
      WRITE(6,103) UL
      DO 9 I=1,M
    9 BB(I)=UL(I+1)
      SUM=0.
      DO 10 K=1,N
      CALL MV(A(1,1,K),M,BB,UU)
      DO 11 I=1,M
   11 BB(I)=UU(I)+B(I,K)
      DO 10 I=1,M
      SUM=SUM+ABS((BB(I)-U(K,I))/BB(I))
   10 U(K,I)=BB(I)
      DO 12 I=1,N
   12 WRITE(6,103) UB(I),(U(I,J),J=1,M),UT(I)
      WRITE(6,103) UR
      IF(SUM.GT.1.E-5.AND.ITER.LT.6) GO TO 13
      STOP
      END
```

ITERATION 0

0.0	0.0	0.0	0.0	0.0	0.0	0.0	0.0	0.0
	0.0	0.0	0.0	0.0	0.0	0.0	0.0	0.0
0.0	0.0	0.0	0.0	0.0	0.0	0.0	0.0	0.0
	0.0	0.0	0.0	0.0	0.0	0.0	0.0	1.00000
0.0	0.0	0.0	0.0	0.0	0.0	0.0	0.0	0.0
	0.0	0.0	0.0	0.0	0.0	0.0	0.0	1.00000
0.0	0.0	0.0	0.0	0.0	0.0	0.0	0.0	0.0
	0.0	0.0	0.0	0.0	0.0	0.0	0.0	1.00000
0.0	0.0	0.0	0.0	0.0	0.0	0.0	0.0	0.0
	0.0	0.0	0.0	0.0	0.0	0.0	0.0	1.00000
0.0	0.0	0.0	0.0	0.0	0.0	0.0	0.0	0.0
	0.0	0.0	0.0	0.0	0.0	0.0	0.0	1.00000
0.0	0.0	0.0	0.0	0.0	0.0	0.0	0.0	0.0
	0.0	0.0	0.0	0.0	0.0	0.0	0.0	1.00000
0.0	0.0	0.0	0.0	0.0	0.0	0.0	0.0	0.0
	0.0	0.0	0.0	0.0	0.0	0.0	0.0	1.00000
0.0	0.0	0.0	0.0	0.0	0.0	0.0	0.0	0.0
	0.0	0.0	0.0	0.0	0.0	0.0	0.0	1.00000
0.0	0.0	0.0	0.0	0.0	0.0	0.0	0.0	0.0
	0.0	0.0	0.0	0.0	0.0	0.0	0.0	1.00000
0.0	0.0	0.0	0.0	0.0	0.0	0.0	0.0	0.0
	0.0	0.0	0.0	0.0	0.0	0.0	0.0	1.00000
0.0	0.0	0.0	0.0	0.0	0.0	0.0	0.0	0.0
	0.0	0.0	0.0	0.0	0.0	0.0	0.0	1.00000
0.0	0.0	0.0	0.0	0.0	0.0	0.0	0.0	0.0
	0.0	0.0	0.0	0.0	0.0	0.0	0.0	1.00000
0.0	0.0	0.0	0.0	0.0	0.0	0.0	0.0	0.0
	0.0	0.0	0.0	0.0	0.0	0.0	0.0	1.00000
0.0	0.0	0.0	0.0	0.0	0.0	0.0	0.0	0.0
	0.0	0.0	0.0	0.0	0.0	0.0	0.0	1.00000
0.0	0.0	0.0	0.0	0.0	0.0	0.0	0.0	0.0
	0.0	0.0	0.0	0.0	0.0	0.0	0.0	0.0

ITERATION 1

0.0	0.0	0.0	0.0	0.0	0.0	0.0	0.0	0.0
	0.0	0.0	0.0	0.0	0.0	0.0	0.0	0.0
0.0	0.00429	0.00876	0.01359	0.01898	0.02517	0.03246	0.04124	0.05205
	0.06570	0.08346	0.10757	0.14231	0.19691	0.29386	0.49570	1.00000
0.0	0.00842	0.01717	0.02662	0.03715	0.04924	0.06343	0.08045	0.10127
	0.12728	0.16057	0.20450	0.26476	0.35149	0.48283	0.68895	1.00000
0.0	0.01220	0.02488	0.03856	0.05378	0.07120	0.09157	0.11586	0.14531
	0.18158	0.22704	0.28509	0.36077	0.46144	0.59701	0.77729	1.00000
0.0	0.01550	0.03160	0.04894	0.06822	0.09020	0.11580	0.14612	0.18251
	0.22670	0.28092	0.34804	0.43177	0.53649	0.66647	0.82321	1.00000
0.0	0.01819	0.03708	0.05740	0.07995	0.10558	0.13531	0.17029	0.21191
	0.26180	0.32188	0.39440	0.48179	0.58629	0.70917	0.84907	1.00000
0.0	0.02018	0.04113	0.06365	0.08858	0.11687	0.14956	0.18784	0.23305
	0.28670	0.35042	0.42590	0.51468	0.61772	0.73486	0.86389	1.00000
0.0	0.02140	0.04361	0.06747	0.09386	0.12376	0.15823	0.19846	0.24576
	0.30152	0.36720	0.44412	0.53329	0.63507	0.74864	0.87165	1.00000
0.0	0.02181	0.04445	0.06876	0.09564	0.12607	0.16114	0.20201	0.24999
	0.30644	0.37272	0.45007	0.53932	0.64061	0.75299	0.87407	1.00000
0.0	0.02140	0.04361	0.06747	0.09386	0.12376	0.15823	0.19846	0.24576
	0.30152	0.36720	0.44412	0.53330	0.63507	0.74864	0.87165	1.00000
0.0	0.02018	0.04113	0.06365	0.08858	0.11687	0.14956	0.18784	0.23305
	0.28670	0.35042	0.42591	0.51468	0.61773	0.73486	0.86389	1.00000
0.0	0.01819	0.03708	0.05740	0.07995	0.10558	0.13531	0.17030	0.21192
	0.26180	0.32189	0.39441	0.48179	0.58630	0.70918	0.84907	1.00000
0.0	0.01550	0.03160	0.04895	0.06822	0.09020	0.11580	0.14612	0.18251
	0.22670	0.28092	0.34805	0.43178	0.53650	0.66647	0.82321	1.00000
0.0	0.01220	0.02488	0.03856	0.05378	0.07120	0.09157	0.11586	0.14531
	0.18158	0.22704	0.28509	0.36077	0.46144	0.59701	0.77729	1.00000
0.0	0.00842	0.01717	0.02662	0.03715	0.04924	0.06343	0.08045	0.10127
	0.12728	0.16057	0.20450	0.26477	0.35149	0.48283	0.68896	1.00000
0.0	0.00429	0.00876	0.01359	0.01898	0.02517	0.03246	0.04124	0.05205
	0.06570	0.08346	0.10757	0.14231	0.19692	0.29386	0.49570	1.00000
0.0	0.0	0.0	0.0	0.0	0.0	0.0	0.0	0.0
	0.0	0.0	0.0	0.0	0.0	0.0	0.0	0.0

ITERATION 2

0.0	0.0	0.0	0.0	0.0	0.0	0.0	0.0	0.0	0.0	0.0	0.0	0.0	0.0	0.0	0.0	0.0
0.0	0.00418	0.00852	0.01322	0.01847	0.02451	0.03164	0.04024	0.05086	0.06431	0.08187	0.10581	0.14042	0.19498	0.29206	0.49436	1.00000
0.0	0.00818	0.01670	0.02589	0.03615	0.04795	0.06180	0.07846	0.09890	0.12451	0.15741	0.20097	0.26097	0.34760	0.47921	0.68634	1.00000
0.0	0.01186	0.02419	0.03749	0.05231	0.06929	0.08918	0.11294	0.14181	0.17750	0.22236	0.27987	0.35515	0.45572	0.59176	0.77361	1.00000
0.0	0.01506	0.03072	0.04758	0.06633	0.08774	0.11272	0.14235	0.17800	0.22142	0.27486	0.34130	0.42454	0.52918	0.65986	0.81870	1.00000
0.0	0.01768	0.03603	0.05579	0.07771	0.10267	0.13165	0.16581	0.20655	0.25551	0.31468	0.38638	0.47321	0.57770	0.70150	0.84394	1.00000
0.0	0.01961	0.03996	0.06184	0.08609	0.11361	0.14546	0.18282	0.22703	0.27964	0.34234	0.41693	0.50512	0.60819	0.72643	0.85833	1.00000
0.0	0.02079	0.04237	0.06555	0.09120	0.12028	0.15386	0.19310	0.23933	0.29399	0.35857	0.43455	0.52313	0.62497	0.73977	0.86584	1.00000
0.0	0.02119	0.04318	0.06679	0.09292	0.12253	0.15667	0.19654	0.24343	0.29874	0.36392	0.44030	0.52895	0.63033	0.74397	0.86817	1.00000
0.0	0.02079	0.04237	0.06555	0.09120	0.12029	0.15386	0.19310	0.23933	0.29399	0.35857	0.43455	0.52313	0.62498	0.73977	0.86584	1.00000
0.0	0.01961	0.03996	0.06184	0.08609	0.11361	0.14546	0.18282	0.22704	0.27964	0.34234	0.41693	0.50512	0.60820	0.72644	0.85833	1.00000
0.0	0.01768	0.03603	0.05579	0.07771	0.10267	0.13165	0.16582	0.20655	0.25551	0.31468	0.38639	0.47322	0.57770	0.70150	0.84394	1.00000
0.0	0.01506	0.03072	0.04758	0.06633	0.08775	0.11272	0.14235	0.17800	0.22142	0.27487	0.34130	0.42454	0.52918	0.65986	0.81870	1.00000
0.0	0.01186	0.02419	0.03749	0.05231	0.06929	0.08918	0.11294	0.14182	0.17750	0.22236	0.27987	0.35515	0.45572	0.59176	0.77361	1.00000
0.0	0.00813	0.01670	0.02589	0.03615	0.04794	0.06180	0.07846	0.09890	0.12451	0.15741	0.20097	0.26097	0.34761	0.47922	0.68634	1.00000
0.0	0.00418	0.00852	0.01322	0.01847	0.02451	0.03164	0.04024	0.05086	0.06431	0.08187	0.10581	0.14042	0.19498	0.29206	0.49436	1.00000
0.0	0.0	0.0	0.0	0.0	0.0	0.0	0.0	0.0	0.0	0.0	0.0	0.0	0.0	0.0	0.0	0.0

ITERATION 3

```
0.0       0.0       0.0       0.0       0.0       0.0       0.0       0.0       0.0
     0.0       0.0       0.0       0.0       0.0       0.0       0.0       0.0
0.0       0.00418   0.00852   0.01322   0.01847   0.02451   0.03164   0.04024   0.05086
     0.06431   0.08187   0.10581   0.14042   0.19493   0.29206   0.49436   1.00000
0.0       0.00818   0.01670   0.02589   0.03615   0.04793   0.06180   0.07846   0.09890
     0.12451   0.15741   0.20097   0.26097   0.34760   0.47922   0.68634   1.00000
0.0       0.01186   0.02419   0.03749   0.05231   0.06929   0.08918   0.11294   0.14181
     0.17750   0.22236   0.27987   0.35515   0.45572   0.59176   0.77361   1.00000
0.0       0.01506   0.03072   0.04758   0.06633   0.08774   0.11272   0.14235   0.17800
     0.22142   0.27486   0.34130   0.42454   0.52918   0.65986   0.81870   1.00000
0.0       0.01768   0.03603   0.05579   0.07771   0.10267   0.13165   0.16581   0.20655
     0.25551   0.31468   0.38638   0.47321   0.57770   0.70150   0.84394   1.00000
0.0       0.01961   0.03996   0.06184   0.08608   0.11361   0.14546   0.18282   0.22703
     0.27964   0.34234   0.41693   0.50512   0.60819   0.72643   0.85833   1.00000
0.0       0.02079   0.04237   0.06555   0.09120   0.12028   0.15386   0.19310   0.23933
     0.29399   0.35857   0.43455   0.52313   0.62497   0.73977   0.86584   1.00000
0.0       0.02119   0.04318   0.06679   0.09292   0.12252   0.15667   0.19654   0.24343
     0.29874   0.36392   0.44030   0.52895   0.63033   0.74397   0.86817   1.00000
0.0       0.02079   0.04237   0.06555   0.09120   0.12028   0.15386   0.19310   0.23933
     0.29399   0.35857   0.43455   0.52313   0.62497   0.73977   0.86584   1.00000
0.0       0.01961   0.03996   0.06184   0.08609   0.11361   0.14546   0.18282   0.22703
     0.27964   0.34234   0.41693   0.50512   0.60819   0.72644   0.85833   1.00000
0.0       0.01768   0.03603   0.05579   0.07771   0.10267   0.13165   0.16581   0.20655
     0.25551   0.31468   0.38638   0.47322   0.57770   0.70150   0.84394   1.00000
0.0       0.01506   0.03072   0.04758   0.06633   0.08774   0.11272   0.14235   0.17800
     0.22142   0.27487   0.34130   0.42454   0.52918   0.65986   0.81870   1.00000
0.0       0.01186   0.02419   0.03749   0.05231   0.06929   0.08918   0.11294   0.14182
     0.17750   0.22236   0.27987   0.35515   0.45572   0.59176   0.77361   1.00000
0.0       0.00818   0.01670   0.02589   0.03615   0.04793   0.06180   0.07846   0.09890
     0.12451   0.15741   0.20097   0.26097   0.34761   0.47922   0.68634   1.00000
0.0       0.00418   0.00852   0.01322   0.01847   0.02451   0.03164   0.04024   0.05086
     0.06431   0.08187   0.10581   0.14042   0.19498   0.29206   0.49436   1.00000
0.0       0.0       0.0       0.0       0.0       0.0       0.0       0.0       0.0
     0.0       0.0       0.0       0.0       0.0       0.0       0.0       0.0
```

Author Index

Subject Index

Mathematics in Science and Engineering

A Series of Monographs and Textbooks

Edited by RICHARD BELLMAN, ***University of Southern California***

23. A. Halanay. Differential Equations: Stability, Oscillations, Time Lags. 1966

24. M. N. Oguztoreli. Time-Lag Control Systems. 1966

25. D. Sworder. Optimal Adaptive Control Systems. 1966

26. M. Ash. Optimal Shutdown Control of Nuclear Reactors. 1966

27. D. N. Chorafas. Control System Functions and Programming Approaches (In Two Volumes). 1966

28. N. P. Erugin. Linear Systems of Ordinary Differential Equations. 1966

29. S. Marcus. Algebraic Linguistics; Analytical Models. 1967

30. A. M. Liapunov. Stability of Motion. 1966

31. G. Leitmann (ed.). Topics in Optimization. 1967

32. M. Aoki. Optimization of Stochastic Systems. 1967

33. H. J. Kushner. Stochastic Stability and control. 1967

34. M. Urabe. Nonlinear Autonomous Oscillations. 1967

35. F. Calogero. Variable Phase Approach to Potential Scattering. 1967

36. A. Kaufmann. Graphs, Dynamic Programming, and Finite Games. 1967

37. A. Kaufmann and R. Cruon. Dynamic Programming: Sequential Scientific Management. 1967

38. J. H. Ahlberg, E. N. Nilson, and J. L. Walsh. The Theory of Splines and Their Applications. 1967

39. Y. Sawaragi, Y. Sunahara, and T. Nakamizo. Statistical Decision Theory in Adaptive Control Systems. 1967

40. R. Bellman. Introduction to the Mathematical Theory of Control Processes, Volume I. 1967; Volume II. 1971 (Volume III in preparation)

41. E. S. Lee. Quasilinearization and Invariant Imbedding. 1968

42. W. Ames. Nonlinear Ordinary Differential Equations in Transport Processes. 1968

43. W. Miller, Jr. Lie Theory and Special Functions. 1968

44. P. B. Bailey, L. F. Shampine, and P. E. Waltman. Nonlinear Two Point Boundary Value Problems. 1968

45. Iu. P. Petrov. Variational Methods in Optimum Control Theory. 1968

46. O. A. Ladyzhenskaya and N. N. Ural't-seva. Linear and Quasilinear Elliptic Equations. 1968

47. A. Kaufmann and R. Faure. Introduction to Operations Research. 1968

48. C. A. Swanson. Comparison and Oscillation Theory of Linear Differential Equations. 1968

49. R. Hermann. Differential Geometry and the Calculus of Variations. 1968

50. N. K. Jaiswal. Priority Queues. 1968

51. H. Nikaido. Convex Structures and Economic Theory. 1968

52. K. S. Fu. Sequential Methods in Pattern Recognition and Machine Learning. 1968

53. Y. L. Luke. The Special Functions and Their Approximations (In Two Volumes). 1969

54. R. P. Gilbert. Function Theoretic Methods in Partial Differential Equations. 1969

55. V. Lakshmikantham and S. Leela. Differential and Integral Inequalities (In Two Volumes). 1969

56. S. H. Hermes and J. P. LaSalle. Functional Analysis and Time Optimal Control. 1969

57. M. Iri. Network Flow, Transportation, and Scheduling: Theory and Algorithms. 1969

58. A. Blaquiere, F. Gerard, and G. Leitmann. Quantitative and Qualitative Games. 1969

59. P. L. Falb and J. L. de Jong. Successive Approximation Methods in Control and Oscillation Theory. 1969

60. G. Rosen. Formulations of Classical and Quantum Dynamical Theory. 1969

61. R. Bellman. Methods of Nonlinear Analysis, Volume I. 1970

62. R. Bellman, K. L. Cooke, and J. A. Lockett. Algorithms, Graphs, and Computers. 1970

63. E. J. Beltrami. An Algorithmic Approach to Nonlinear Analysis and Optimization. 1970

64. A. H. Jazwinski. Stochastic Processes and Filtering Theory. 1970

65. P. Dyer and S. R. McReynolds. The Computation and Theory of Optimal Control. 1970

66. J. M. Mendel and K. S. Fu (eds.). Adaptive, Learning, and Pattern Recognition Systems: Theory and Applications. 1970

67. C. Derman. Finite State Markovian Decision Processes. 1970

68. M. Mesarovic, D. Macko, and Y. Takahara. Theory of Hierarchial Multilevel Systems. 1970

69. H. H. Happ. Diakoptics and Networks. 1971

70. Karl Astrom. Introduction to Stochastic Control Theory. 1970

71. G. A. Baker, Jr. and J. L. Gammel (eds.). The Padé Approximant in Theoretical Physics. 1970

72. C. Berge. Principles of Combinatorics. 1971

73. Ya. Z. Tsypkin. Adaptation and Learning in Automatic Systems. 1971

74. Leon Lapidus and John H. Seinfeld. Numerical Solution of Ordinary Differential Equations. 1971

75. L. Mirsky. Transversal Theory. 1971

76. Harold Greenberg. Integer Programming. 1971

77. E. Polak. Computational Methods in Optimization: A Unified Approach. 1971

78. Thomas G. Windeknecht. General Dynamical Processes: A Mathematical Introduction. 1971

79. M. A. Aiserman, L. A. Gusev, L. I. Rozonoer, I. M. Smirnova, and A. A. Tal'. Logic, Automata, and Algorithms. 1971

80. Andrew P. Sage and James L. Melsa. System Identification. 1971

81. R. Boudarel, J. Delmas, and P. Guichet. Dynamic Programming and Its Application to Optimal Control. 1971

82. F. V. Atkinson. Multiparameter Eigenvalue Problems, Volume I. 1972

83. William S. Meisel. Computer-Oriented Approaches to Pattern Recognition. 1972

84. M. Frank Norman. Markov Processes and Learning Models. 1972

85. G. E. Ladas and V. Lakshmikantham. Differential Equations in Abstract Spaces. 1972

86. William T. Reid. Riccati Differential Equations. 1972

87. Bruce A. Finlayson. The Method of Weighted Residuals and Variational Principles: With Application to Fluid Mechanics, Heat and Mass Transfer. 1972

88. Edward Angel and Richard Bellman. Dynamic Programming and Partial Differential Equations. 1972

In preparation

Alexander Weinstein and William Stenger. Methods of Intermediate Problems for Eigenvalues: Theory and Ramifications

G. Arthur Mihram. Simulation: Statistical Foundations and Methodology

Renfrey B. Potts and Robert M. Oliver. Flows in Transportation Networks

Umberto Bertele and Franchesco Brioschi. Nonserial Dynamic Programming

H. Melvin Lieberstein. Theory of Partial Differential Equations

H. M. Nussenzveig. Causality and Dispersion Relations